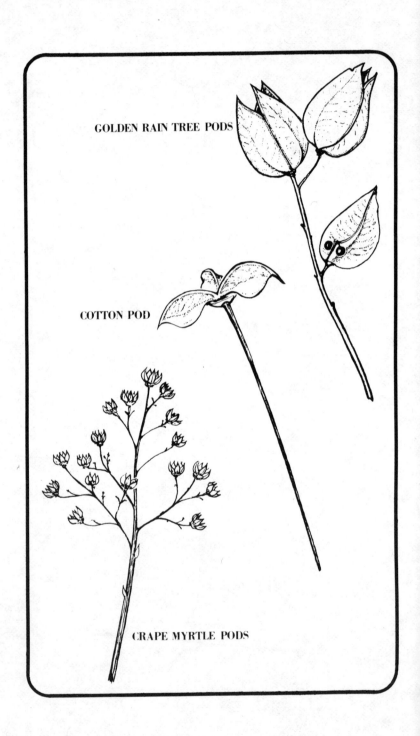

GOLDEN RAIN TREE PODS

COTTON POD

CRAPE MYRTLE PODS

DRIED GRASSES, GRAINS, GOURDS, PODS AND CONES

by

Leonard Karel

The Scarecrow Press, Inc.
Metuchen, N. J.   1975

Credits:

Drawings by Rothenberg Advertising
Photographs by Martin L. Karel
Manuscript typing by Eva (Mrs. Louis) Rothenberg

Library of Congress Cataloging in Publication Data

Karel, Leonard, 1912–
    Dried grasses, grains, gourds, pods, and cones.

    Bibliography: p.
    Includes index.
    1. Plants--Drying. I. Title.
SB447.K293   745.92   74-31178
ISBN 0-8108-0792-0

745.92
K18

Copyright 1975 by Leonard Karel

PROLOGUE

And the earth brought forth grass, and herb yielding seed after his kind, and the tree yielding fruit, whose seed was in itself after his kind. (Genesis: 1, 12)

And shewed them the fruit of the land ... (Numbers: 13, 26)

The precious fruits brought forth by the sun ... (Deuteronomy: 33, 14)

Better than gold, yea, than fine gold. (Proverbs: 8, 19)

CONTENTS

ILLUSTRATIONS

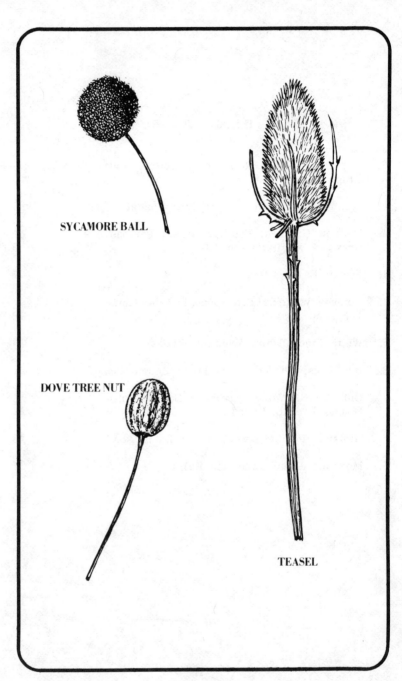

SYCAMORE BALL

DOVE TREE NUT

TEASEL

Figure 1

HISTORY

Ever since Adam and Eve introduced the fig leaf to the world of high fashion, man's commitment to plants has been total and indissoluble. Records of ancient civilizations amply confirm this intimate association.

Inspired by the beauty of the plant world, man has searched from antiquity to the present for ways of preserving and enjoying this beauty beyond its season. Efforts to retain and to imitate the loveliness of the flower, the gracefulness of the leaf, and the architecture of the fruit have been continual and expanding.

We may reasonably assume that our ancestors first directed their talents to the preservation of flowers. Gradually, however, they must have become aware of the pleasing qualities of many other botanical elements--grasses and grains, nuts and gourds, drupes and berries, pods and cones, lichens and leaves, bark and branches, shapely vines, and weathered deadwood.

Thus, 3,000 years ago, in the Swiss neolithic period, a craftsman decorated Cortaillod pots with birch-bark patterns glued to the outside with wood pitch. Similar patterns ornamented the grips of combs. [Vogt]

In 1727, Francis Xavier d'Entrecolles wrote from Peking that the Chinese manufactured flowers from the pith of Aralia papyrifera. [Bretschneider]

Ancient Egyptians entombed their deceased with a

1

remarkable variety of plants.   Flowers and leaves occurred
with frequency.   The finding of wheat (<u>Triticum vulgare</u>) has
been common.   A tomb of the 5th Dynasty (2,494-2,345 B.C.)
held barley (<u>Hordeum vulgare</u>).   The tomb of Dra-Abu-Negga,
of the 12th Dynasty (1,991-1,786 B.C.), contained the common bottle gourd, calabash (<u>Lagenaria vulgaris</u>), and small,
well-preserved, characteristic pine cones, along with lichens
and juniper berries.   The pine cones, lichens, and berries
had probably been imported. [Schweinfurth]

In the discoveries at Dar-el-Bahari (Thebes) in 1881,
archaeologists found that tombs of the 22nd Dynasty (945-730
B.C.) held ginger grass--also called camelgrass--(<u>Andropogon
laniger</u>), eragrostis (<u>Eragrostis cyanosuroides</u>), and lichens
entwined with tendrils of cucurbits.   [According to L. H.
Bailey, gourds have probably been grown since prehistoric
times.]

Egyptians viewed death as a change to a new existence
in which the departed would require the kinds of things to
which he had become accustomed in life.   The greater the
social position of the deceased, the higher the honors ac-
corded him in life, and the more varied his tastes, the more
abundant and elaborate the materials entombed with his mortal
remains.

Tombs of royalty, aristocracy, and other persons of
high social rank have yielded many botanical riches.   To
what extent some of these natural materials represent purely
ornamental items bearing a particular meaning and importance
to the deceased and to the bereaved, we can only conjecture.
To what extent some of these materials may have been consciously preserved, we can only wonder.   Perhaps as the
deceased was readied for his new, everlasting existence,
some durable flowers, seed vessels, and other plant sections

carried a special significance related to their everlasting
qualities.

The use of non-botanical material to imitate the botanical goes back many years. "Crassus the Rich [King
Croessus, 560-546 B.C.] was the first to make <u>artificial</u>
(underscore mine) leaves of silver or gold, giving chaplets of
them as prizes at his games, to which were also added ribbons." [Pliny--Book **XXI**:   III and **IV**, page 165]

The general manufacture of artificial flowers, leaves,
and fruit occurred around 375 B.C. "Floral chaplets being
now fashionable, it was not long before there appeared what
are called Egyptian chaplets, and then winter ones, made
from dyed flakes of horn at the season when earth refuses
flowers."

"Marcus Varro [about 110-56 B.C.] records that he
knew at Rome an artist named Possis who made fruit and
grapes in such a way that nobody could tell by sight from the
real things." [Pliny--Book **XXXV**:   XLV 156, page 375]

Herbalists of the fifteenth and sixteenth centuries "val-
ued certain grasses not only for food, but also for ornament,"
according to <u>Gerard's Herball, 1633,</u> * cited by Agnes Arber.
In the same herbal, "we get a glimpse of grasses used in
the decoration of living rooms, for we are told that there is
a plumy grass called "windlestrawes," "with which we in London do usually adorn our chimneys in Sommer time: and
we commonly call the bundle of it handsomely made up for
our use, by name of Bents."

"The variegated form of <u>Phalaris arundinacea L.</u> must
have been long a treasured garden plant. It is noticed in

---

*Arber points out that the text of none of three herbals--one
published in 1542; a second in 1551 and 1562; and the third
in 1633 (Gerard's)--is strikingly original.

Gerard's herbal under the name of "Ladies-lace Grasse."
It is described as having leaves "with many white vaines or
ribs, and silver streaks running along through the midst of
the leaves, fashioning the same like to laces or ribbons woven
of white and greene silke, very beautiful and faire to behold."

In her chapter on "Ikebana: Japanese Flower Arrangement," Julia Berrall noted: "The pine, the bamboo, and the
early plum are a favorite combination in Japan, just as they
are in China; and another group of plant materials that one
sees frequently is that of the so-called 'seven flowers of
autumn,' a combination which includes grasses as well as
flowers."

During the Victorian period, according to Berrall, "It
became very fashionable to bring Nature indoors, as Godey's*
tells us, by gathering 'the various grasses of the season and
allowing them to dry in a dark room before tastefully arranging them in a vase or pretty basket. Wheat, oats, and rye,
as well as the various tall grasses to be found every place in
the country, make beautiful ornaments for the parlor.'"

Berrall has also told us: "Striped ribbon grass is
often seen in old flower prints, and feathery grasses and dock
seem often to have been included in bunches of flowers."

Speaking of ornamental grasses, Kresken wrote (in
1887): "The variety of grasses that can be made available
for ornamentation is very great; but I will here only name a
portion of them, and such as I consider best for the purpose.
For general information, I will give the botanical as well as
the English names of most of the following grasses:

---

*Godey's Lady's Book--a monthly magazine published around
1855.

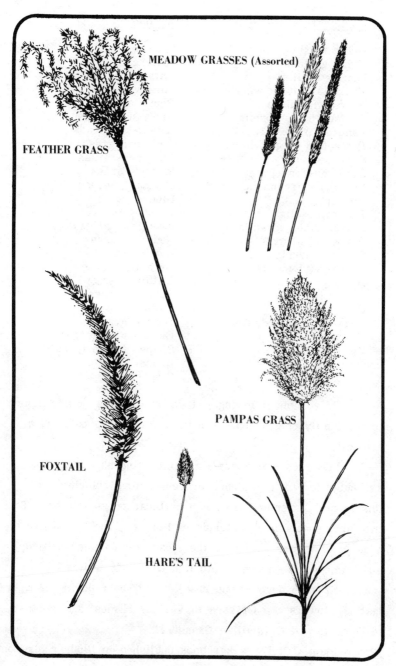

FEATHER GRASS

MEADOW GRASSES (Assorted)

FOXTAIL

HARE'S TAIL

PAMPAS GRASS

Figure 2

Botanical	English
Briza maxima	Quaking - grass
Briza geniculata	Shaking - grass
Briza gracilis	Large shaking - grass
Phalaria canariensis	Canary - grass
Agrosti nebulosa	Bent - grass
Bromus briza formis	Drooping - grass
Lagurus ovatus	Rabbit-tail grass
Penisetum longistylum	Long-pen grass
Stipa pennata	Feather - grass
Eryanthus ravennae	Sugar - grass
Gynerium argenteum	Pampas - grass
Poa pratensis	Blue - grass
Poa nemoralis	Wood - grass
Anthoxanthum gracile	Sweet vernal grass
Agrostis vulgaris	Redtop - grass
Lolium perenis	Rye - grass
Dactillis glomerata	Orchard - grass
Briza pyrum siculum	Glitter - grass
Cloris barbata	Beard - grass
Coix lacryma	Job's tears
Eragrostis cylindriflora	Love - grass
Zea caragua	Giant maize (corn)
Zea japonica	Striped maize (corn)
	Broom corn
	Sorghum

"Also, the following well-known cereals, which belong to the family of grasses: Wheat, rye, barley, oats, rice, and millet.

"Grasses will assume specially fine shades if they are subjected to the sulphur process before coloring. If this is not convenient, they can be bleached as follows: Take two table-spoonfuls of chloride of lime and one of vinegar to a quart of water. Pour off the clear portion when settled, and immerse the grasses into it until nearly white."

Kresken devotes one chapter to "The Coloring of Everlasting Flowers and Grasses in Various Shades" and another to "Bronze and Crystallize Grasses."

Plant Culture, a handbook published in 1900, refers on

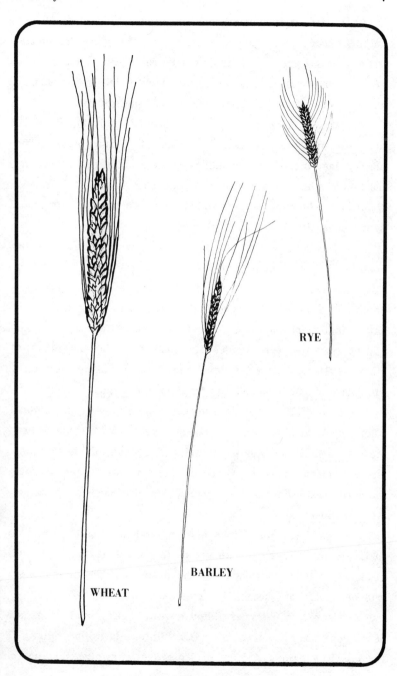

WHEAT

BARLEY

RYE

Figure 3

pages 92 and 93 to a number of ornamental grasses. The
author, George W. Oliver, singles out "Erianthus Ravennae":
"The growths are stout, ending with very ornamental flowers, which if taken in a young state and dried in the sun,
are quite as showy as those of the Pampas plumes."

    Writing on the Pampas Grasses (Cortaderia, Stapf.)
in Flora and Sylva, Volume III, 1905, Otto Stapf said, "The
plumes last therefore for many weeks even in the open, until
the rough winds of autumn break the shafts, and it is this
endurance which makes the dried heads so valuable for in-
door decoration.

    "I have said that the Pampas-grass first came to this
country [England] more than half a century ago. It was in
1843 that David Moore of Glasnevin received seed from
James Tweedie--at that time collecting in the Argentine. Its
actual discoverer however was Frederick Sellow of Potsdam
... he travelled with Prince Maximilian von Wied (1815-1817)
and in 1819 with Van Olfers of the Prussian Legation. On
this occasion he reached Uruguay (or the Banda Oriental as
it was then called) and discovered the Pampas-grass somewhere near Montevideo.... Sellow was drowned soon after
his return to Rio de Janeiro, leaving no account of his travels.
His collections however reached Berlin in safety, and from
them (in 1825) Sprengel described our plant under the name
Arundo dioeca.

    "... Vast plantations of Pampas-grass have sprung up
in California, where it was grown for profit as long ago as
1874. The crop of 1889 from the Santa Barbara plantations
was estimated at one million plumes."

    George W. Carver listed several ornamental grasses
of Macon County, Alabama. Of "Uniola latifolia (Spike
Grass)," he said, "A very pretty grass, attaining a height of

two to three feet. It is a rapid grower and exceedingly attractive in flower and fruit. As a dried grass it is one of
the best. It thrives best in low, wet, partially shaded
places. "

W. J. Beal, in his contribution to Bailey's Standard
Cyclopedia of Horticulture, 1944, page 1394, wrote: "Grasses
are not so much employed for ornamenting homes as their
merits warrant.... For table decoration nothing is better
than the elegant, airy panicles of large numbers of wild
grasses, such as species of Poa, Koeleria, Sphenopholis, ◆
Panicum, Paspalum, Eragrostis, Muhlenbergia, Bromus,
Festuca, Agrostis, Deschampsia, Uniola, Cinna latifolia.
For large halls and exhibitions, nothing surpasses sheaves
of wheat, barley, rice, oats or any of the wild grasses. For
decoration, grasses should be cut before ripe, dried in the
dark in an upright position, and may be used in that condi-
tion or dyed or bleached. "

Volume II, page 922, of the Royal Horticultural Society's 1956 edition of Dictionary of Gardening states: "Various grasses find a place in gardens for purposes other than
their value as lawn plants. Several besides the Bamboos
are of noble proportions and graceful habit and are very ornamental in large borders or as isolated lawn plants; some
are valuable on the rock-garden or for their foliage in the
border, or for edgings; some have a double value: in the
border for their graceful spikes or panicles which, being
dried, may also be used with everlasting flowers, &, for in-
door decoration in winter. For this purpose they should be
cut on a fine day before the seeds ripen and gradually dried
in a cool place....

"The Bamboos are distributed over the following gen-
era: Arundinaria, Bambusa, Dendrocalamus (tender), Phyl-

lostachys, Sasa, Shibataea, and though not a true Bamboo, Chusquea may also be included among them for its similar garden value.

"Other tall ornamental grasses for lawn or bank plants and other more or less isolated positions will be found under Arundo, Cortaderia, Elymus, Erianthus, Gynerium, Saccharum, Zea.

"Somewhat dwarfer plants more suitable for borders, or in some instances for pots, or for the water-side, are under Agrostis, Aira, Andropogon, Arrhenatherum, Asperella, Briza, Bromus, Calamovilfa, Chloris, Coix, Cymbopogon, Dactylis, Deschampsia, Desmazeria, Eleusine, Eragrostis, Festuca, Glyceria, Hierochloe, Holcus, Hordeum, Lagurus, Lamarckia, Melica, Miscanthus, Molinia, Panicum, Pennisetum, Phalaris, Phragmites, Poa, Stipa, Trichloris, Tricholaena, Uniola. Several of these produce inflorescences useful for winter bouquets, others have variegated forms less vigorous than the normal form and more amenable to use in the garden. "

In her excellent history of flower arranging, Julia Berrall reported that in colonial and eighteenth century America, "Dried flowers were quite commonly gathered and bunched for the winter months. Native materials which we definitely know to have been used included pearly everlasting, sea lavender, and bitter-sweet; sumac, alder, and cattails; ... berries and seed pods could also have been used. [underscore, ours]. Gardens contributed globe amaranth, honesty [lunaria pods], and strawflowers. "

Marguerite Yates wrote that the Victorian period saw the decorative use of "dried and artifically colored seeds of practically every form of vegetation. " The list includes watermelon, pumpkin, lima beans, ordinary beans, coffee, peanuts, kidney beans, wheat, dates, sunflowers, hollyhock, peas, rice, and muskmelon. Some (the darker ones) were used in their natural colors and others were either dyed or painted....

"Accessory materials were glue, cardboard, wire, string, thread, varnish, dyes and oil paint, green crepe paper, etc."

The Victorians made flowers; modern-day practices have extended the art to portraits, scenes, and abstracts.

Carver referred to Brunnichia Cirrhosa (Climbing smartweed) as a vine of rare beauty. He said that the flowers were followed by curiously winged seed capsules, making it very ornamental.

CONTEMPORARY PRACTICES

The use of dried plant materials for decorative purposes has reached such proportions that conceiving of a new use becomes difficult. Florist shops, department stores, art galleries, food stores, crafts and hobbies stores, souvenir shops, decorators' lounges, and many more places now sell or display a wide range of items, many of which have been imported from various parts of the world. In travels across the United States, I have seen not only pods, cones, grasses, gourds, and grains, but also branches, vines, husks, tassels, cobs, awns, citrus fruits, berries, seed mosaics, lichens, fungi, mosses, bark, cork, cactus, driftwood, and deadwood. Even with this long list, I feel that I have probably forgotten to mention some other things.

The art of creating pine cone topiary was at one time quite popular and still claims adherents. Combinations of pods with branches such as those of Scotch broom, or with everlastings such as cat-tails, yarrow, goldenrod, or statice have been steadily growing.

Sweet gum balls still enjoy considerable usage and can be purchased from mail order houses. Gilted or silvered or lacquered, the balls can be fashioned into bells or into containers holding floral arrangements. Colored locust pods can be strung to provide a festive overhead decoration at Hallow'een. Burs of chestnuts provide a rich color which blend well with golds and yellows. Rose hips have uses other than supplying vitamin C.

Flowers, leaves, grasses, and grains illuminate book-
marks, letterheads, greeting cards, note paper and place
cards. White or yellow messages can be written on glycer-
inized magnolia leaves. Wreaths incorporate pods, cones,
fruits, nuts, and berries.

Some berries persist for months, without any attempts
to preserve them. A particular favorite of florists is bitter-
sweet (Celastrus scandens).

The ornamental life of berries can be extended by
preventing or retarding water loss. The latter can be accomplished by dipping berries in a mixture of one part shel-
lac and two parts alcohol, or by coating them with liquefied
wax, or by spraying them with a plastic spray. Other types
of coating will suggest themselves to you; the ones given
here serve merely as examples.

Among frequently cited berries you will find these:

```
Barberry (Japanese) - Berberis thunbergii
Bayberry - Myrica pensylvanica
Beautyberry - Callicarpa japonica
Bittersweet - Celastrus scandens
Burning Bush (Euonymus) - Euonymus europaeus
Castor Bean - Ricinus communis (Very poisonous)
Euonymus (Burning Bush) - Euonymus europaeus
Hawthorns - Crataegus species
   (American Hawthorn - Crataegus coccinea)
Holly (American Holly) - Ilex opaca
Honeysuckle - Lonicera species
   (Japanese Honeysuckle - Lonicera japonica)
Privet - Ligustrum vulgare
Pyracantha - Pyracantha species
Snowberry (Waxberry) - Symphoricarpos albus
Sumac (Staghorn Sumac) - Rhus typhina
Waxberry (Snowberry) - Symphoricarpos albus
Winterberry - Ilex verticillata
```

Because they glycerinize so readily, the feathery awns
of clematis deserve special attention. The awns, which turn

from silky masses of shimmering, silvery green to tannish
beard-like masses, were known many years ago as "old man's
beard." One look will tell you why they acquired this appel-
lation.

The individual "petals" of pine cones--especially white
pine--can be painted and fashioned into colorful wheels, tri-
angles, and other geometric forms. Sprays of small capsules or pods can be touched with acrylic colors to give them
an elegance lacking in their natural states. Nuts of the dove-
tree can be colored and mounted to look like so many vari-
colored lollipops. My wife has most effectively used colored
pine-cone flowers, colored dove-tree nuts, and colored cotton
pods to surround the base of a tall, clear-glass bowl in which
lighted circular, colored candles were floating.

In selecting a plant for the ornamental potential of its
parts, look beyond the immediate environment of the plant.
A drab-looking cluster of pods surrounded by withered leaves
and residing in a field of unsightly weeds can often be trans-
formed into a thing of beauty merely by the manner of its
display. Against a new background, joined by appropriate ac-
cessories, and perhaps clothed in a different color, it may
assume an unbelievable richness and interest.

Reported species and varieties of plants far exceed
300,000. Obviously, therefore, plant elements suitable for
decorative uses number in the many thousands. At best
most of us can become personally acquainted with relatively
few. Whatever the extent of our familiarity, much can be
done with little.

Materials that I have overlooked, underestimated, dis-
dained, or discarded one year, I have acquired or re-ex-
amined a following year to good advantage. Sometimes a
specimen has been used for texture, sometimes for size and

accent, and sometimes for color or form. The use of household bleach alone has introduced an astonishing number of variations, even among members of the same species.

The tables citing pods, cones, grasses, and other plant components provide only a limited view of the opportunities available for ornamental innovation. The tables, in other words, should be regarded as indicative rather than exhaustive. They have been constructed to serve as guides to materials already used by someone to create a display, and to materials which can be adapted to decorative purposes.

On one of my hunting expeditions, I found a very striking mullein stalk about three feet in length. The capsule-laden branches numbered six; each branch differed in length from the others--in some instances by two or three inches, and in others by as much as six inches. The lowermost branch left the stalk at a point approximately half the distance from the bottom of the stalk. The branches were so spaced that they projected the appearance of a cactus. From some angles, the mullein resembled a candelabrum. Although attractive to me, the mullein by itself elicited no favorable comment from others. Then I carefully draped it in a few curvaceous sprays of the silvery-gray, dried foliage of artemisia and inserted the stalk into a container fourteen inches tall and three inches wide. Remarks on how beautiful the arrangement looked came often--and without solicitation.

Knowing of my interest in natural materials, a friend brought me seven gray-green, velvety, fascinatingly-twisted Chinese wisteria-pod halves. With the assistance of glue, floral tape, and floral wires, I attached the pods to wires of varying lengths. Immediately, the array assumed a character altogether lacking in the pods found scattered under the

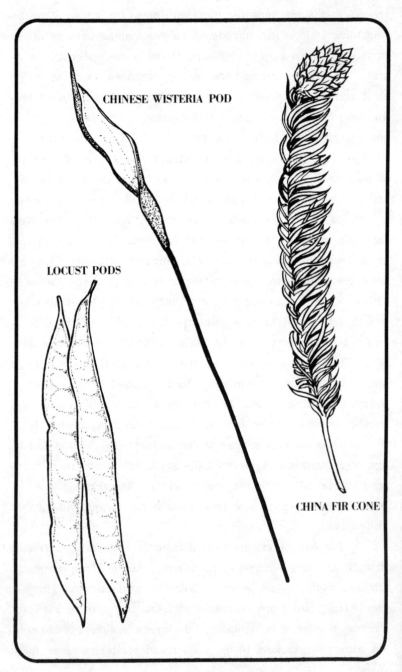

Figure 4

trees. Painting some pod exteriors with diluted orange acrylic paint and other exteriors with yellow paint further enhanced pod-appeal. I introduced an extra fillip by painting one interior with burnt umber and a second with yellow paint. An attempt to bleach one of these pods with undiluted bleach proved basically unsuccessful, and after this pod had dried, I painted it.

My daughter-in-law presented me with about 4-inch lengths of brown well-denuded stalks topped with a spruce cone. Not knowing exactly how best to handle this item, it occurred to me that bleaching the cone might result in a worthwhile object. The cone did not turn out at all well, but the stem did. It came out pure white, beautifully structured and clearly displaying its many knob-like protusions. With a little patience, I was able to attach this exciting objet d'art to a wire stem and immediately used it in a colored pine-cone flower arrangement to bring out the colors by contrast. I found, too, that the protusions could readily be colored to produce a striking effect against their white backgrounds.

Petal-like extensions, which I shall call pine-cone petals, radiate somewhat spirally in layers from the central axis or stem of cones, like those of white pine. Completely severing the stem below, or both above and below, a layer of pine-cone petals produces a form containing several petals. These petals surround the stem much in the manner of petals on real flowers. Removal of superfluous pine-cone petals leaves an attractive flower which can, if one chooses, be modified by bleaching, painting, or dyeing.

Pine-cone flowers may be fashioned in another way. Remove the individual petals from the cone. Using glue, attach these petals to the lower end of the naturally dried center of a coneflower, for example. These manufactured flow-

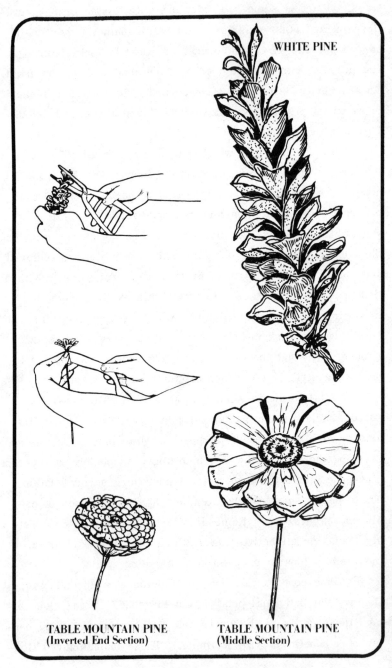

WHITE PINE

TABLE MOUNTAIN PINE
(Inverted End Section)

TABLE MOUNTAIN PINE
(Middle Section)

Figure 5

ers can be left on their natural coneflower stems or mounted on artificial ones.

The most favorable time for working with pine cones happens to be when they are most resinous and sticky. Although solvents like varsol will dissolve the resin, invariably some of the resin sticks to the hands. Furthermore, bathing the cones in varsol before cutting and fashioning them tends to dry the cones as well as the hands, and may result in the closing of the cones. What should one do to overcome the stickiness? How can one easily remove resin from the hands?

I fashion my flowers from the un-varsoled cone and when I can't stand the stickiness any more, I "wash" my hands thoroughly in shortening (other greases or oils will do), and follow this by washing with soap and water. The method works miraculously well. Resin on the pine-cone petals can be removed by rubbing the resinous areas with a rag dampened with varsol. If your fingers pick up some of the varsol, as they will undoubtedly do, use the "shortening" treatment to overcome the loss of oil from your skin.

You may read that heating cones in an oven provides a simple, easy method of removing resin from the cones. True--but the resin that evaporates from the cones must go somewhere. You guessed it; the resin winds up on oven walls if the oven remains closed during the heating.

Be cautious in the use of water or water-containing materials when working on cones. When the latter get wet, they close. Closed cones will re-open as they dry.

You will as a rule wish to insert a stem into your flowers. In some instances, the stem part of the coneflower cannot be readily penetrated by floral wire; nor can an opening, or channel, be made by pressing a thin nail to the center of the cone stem. In such instances, which occur fre-

quently with the larger, sturdier cones, the channel can be
fashioned with a hand drill or an electric drill, and a small
bit.   To avoid breakage of the cone, the latter must either
be clamped gently or held cautiously.

Should the channel be too large for the wire, glue can
be used to secure the wire.   Enlarging the diameter of the
wire by applying a layer of tape around it before inserting
the end of the wire into the cone may also be used to anchor
wire to cone.   A drop of glue applied at juncture of taped
wire and cone helps to anchor the wire even more securely.

Especially interesting ornaments can be made from
magnolia "pods."   Soften the capsules comprising the "pod"
by soaking the latter in water for several hours.   The length
of required soaking varies with the age and dryness of the
"pod."   Remove the individual capsules by gently twisting
each with long-nose pliers and you will be left with a brown,
unusual, spiny decorative "skeleton" which can be readily
mounted on a wire stem.

Sprays of acorn cups look charming when mounted on
stems.   Because of the hardness of their wood, the natural
stems generally must be fastened with floral tape to the wire
stems.   A spot of glue at the contact points between tape
and natural stem helps a great deal to keep the connection
firm.

Acorn cups vary considerably in size; moreover, there
is a great deal of variation in the number of cups comprising
a cluster.   Whereas clusters provide an element of beauty
missing in single cups, some of the latter take on a special
look when attached to stems.   Like pine-cone flowers, acorn
cups may be painted to give a riot of colors.

The China fir, royal paulownia, varieties of rhododen-
drons and azaleas supply lovely seed vessels which can, with

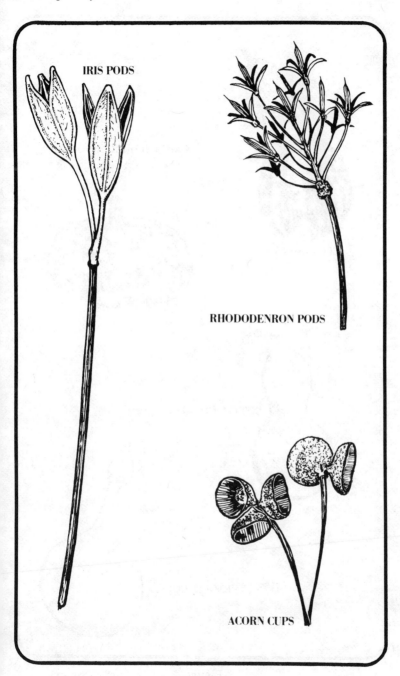

Figure 6

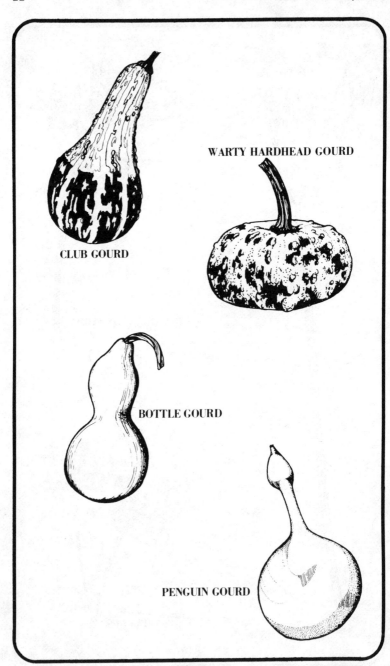

WARTY HARDHEAD GOURD

CLUB GOURD

BOTTLE GOURD

PENGUIN GOURD

Figure 7

little trouble, be mounted or attached to stems. Unfortunately, except for the China fir, these tend to be somewhat fragile.

Hibiscus pods have been very popular when painted a burnt orange on the inside, with the outside left in its natural color. Of course, the outside can be painted too.

Aside from their use as rattles, bird-houses, and containers, gourds have enjoyed ornamental uses for many years now. To dry gourds satisfactorily, follow these steps:

1. Select a perfect gourd that has not been waxed or otherwise treated in such a way as to imprison moisture in the outer rind.

2. Wash the exterior thoroughly with soap and water to remove dirt and molds; then dry it. If you wish, soak the dried gourd for two minutes in a solution of one part of household bleach to three parts of water. This treatment will assist further in ridding the surface of molds. Following the bleach dip, wash the gourd in running water. Now dry it again, and do a good job.

3. Make a small opening at both ends of the gourd; the opening should be about the size of a thin pencil--large enough to allow free circulation of air and the consequent drying of the gourd interior.

4. Place the dried, "opened" gourd in a well-ventilated place with a good exposure to light. Darkness and dampness encourage mold growth. Even though you have hopefully rid the exterior surface of its molds, you may still run afoul of a mold or two which falls from the air and ensconces itself on your gourd's exterior.

5. Be patient. Allow the gourd to dry completely before using it. The drying process may take weeks. As soon as the seeds inside the gourd rattle noisily when you

shake it, the desiccation has been successful.

You may now with impunity wax, spray, or polish your gourd, with assurance that the final product will last a long time. Dependent upon the kind, size, shape, and color, you may wish to section it to serve as a container for a floral arrangement; or, you may choose to drill holes large enough for insertion of flower stems--with the flowers still on them, of course.

BLEACHING

Just how long bleaching has been employed to enhance the beauty of pods and capsules, I do not know. As already mentioned, Kresken, an Ohio florist, bleached flowers, leaves, and grasses almost a century ago.

Bleaching often does more than enhance beauty; bleaching not infrequently creates beauty. The careful, judicious use of bleach opens new vistas. It leads to chromatic and textural modifications which introduce delicate contrasts enriching the accompanying unbleached materials.

Some acorn cups bleach easily; others seem to be overly resistant. Those that bleach completely turn a light tan or beige. The cups of any--bleached or unbleached--can be painted, preferably after first attaching the individual cups or the cluster to wires. The cups handle much more easily when one can hold the wire while painting.

Hibiscus pods bleach to a pure white at times; to an ivory-white at other times.

As mentioned earlier, the knob-like protusions (pulvini) of spruce bleach readily and create an especially pleasing effect against their white background.

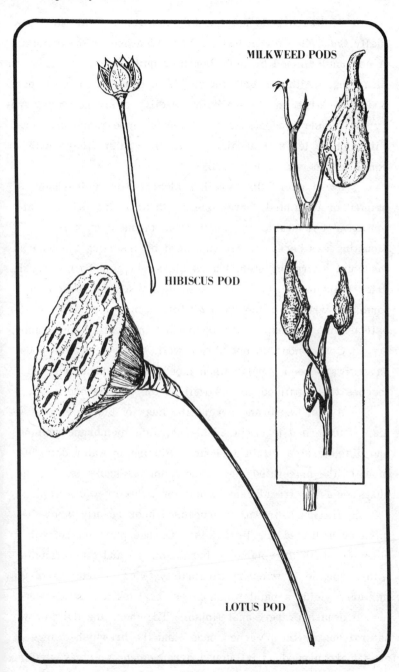

MILKWEED PODS

HIBISCUS POD

LOTUS POD

Figure 8

Milkweed pods take paints beautifully. This is especially true when the pods have been bleached before painting. A pod unblemished by mold becomes quite strikingly white on bleaching. Although both the inside and outside of milkweed pods (the large ones) take paint readily, often the shiny, ivory like inside presents such a beautiful appearance that you will wish to leave it unpainted. Acrylic paint diluted with an equal part of water does well.

Sweet gum balls, whether bleached or unbleached, painted or unpainted, have time and again attracted wide attention and favorable comments when mounted on wire. Mounting can be easily accomplished by inserting the wire at the point where the stem leaves the sweet gum ball. Fresh, crisp balls of the sweet gum tree do not always bleach uniformly. When only the "points" turn white or beige, the ball assumes an especially unique and interesting appearance.

Cotton pods do not bleach well; however, cotton pods have always been popular when mounted; this popularity touches the unpainted pod as well as the painted one.

A thin membrane covers the husk of the royal paulownia. Submersed in undiluted bleach, this membrane whitens and disintegrates within an hour. Rinsing in water detaches most of the macerated membrane from the husk, leaving the husk somewhat translucent except for adherent white areas. The natural dark-brown, unprocessed husk readily takes diluted or undiluted acrylic paints. Orange paint mixed with about one-fifth its volume of burnt umber, and then diluted with a volume of water approximately twice the volume of the mixture, yields a paint which gives the husk a natural-looking, brownish-orange, mat finish. Together, the dark-brown natural husks, the bleached ones, and the brownish-orange ones make a subdued fall-toned arrangement which is striking.

MAGNOLIA POD (with capsules)

MAGNOLIA POD (Capules removed)

SWEETGUM BALL
(Before bleaching)

SWEETGUM BALL
(After bleaching)

Figure 9

Bleached pods, cones, and the like can be easily
painted or tinted.   White, bleached sycamore balls or hibiscus pods, for example, take diluted acrylic paint with ease.

Leafless branches of Eastern hemlock with the cones
still attached to them present a most attractive texture and
form strongly suggestive of a candelabrum.   Bleaching the
"candelabrum" results in somewhat diffused light and dark
areas which suggest a play of lights.

Intact magnolia "pods" bleached for four to six hours
assume a mottled appearance, with the tips of many of the
capsules an off-white.   Occasionally, the entire "pod" bleach-
es to the off-white color.   The brown, skeletal decapsulated
magnolia "pods" when bleached present a striking and unique
appearance.

Cones submersed and kept submersed for six to 12
hours in undiluted household bleach acquire an ivory-like or
a gray-tan color; however, isolated parts of the cone may re-
main unbleached.   This phenomenon enhances rather than de-
tracts from the beauty of the bleached cone.   Although cones
gradually close on contact with water, most will reopen as
they dry.   Heat accelerates the reopening.   In applying heat,
be careful to avoid damaging the cone.

Flowers can be fashioned from the cones either before or after bleaching.   Dried, hard, brittle cones which
are difficult to fashion into flowers can be softened by immersion in water for approximately half an hour; at times,
longer immersion is necessary.   Once prepared, the bleached
cone-flower can be attached to a wire stem, and colored if
one so elects.

ARRANGEMENTS

Principles of arrangement for pods, cones, grasses,
and gourds do not differ from basic principles applied to
flowers. The most basic principle still remains "the arrangement that pleases you. " Just make sure that you like what
you do.

Obviously, techniques and designs applied to bouquets
differ from those applied to pictures or to wreaths. Also,
bouquets in the Victorian style differ in composition from bouquets created in the Ikebana tradition.

As with flowers, you may wish consciously or otherwise to consider some or all of these factors: purpose, area
of display, general background, season of the year, and light
values of the room, hallway, or corner for which you intend
the arrangement. You may wish, also, to consider structure,
texture, color, dimensions, and appearance under different
kinds of light.

In general, you will find a significant difference between pods, cones, grasses, and gourds on the one hand, and
flowers on the other. With occasional exceptions, the former as a class retain their natural color much longer than do
the latter. Additionally, the inherent sturdiness of most pods,
cones, grasses, and gourds far exceeds that of most flowers;
in short, the durability of the former greatly exceeds that of
the latter.

Although the general idea of arrangement can be got-
ten from any pictures of home-made bouquets, I do recom-
mend that you examine publications devoted principally to arrangement. You will, I feel confident, enjoy doing so. (See
the references which follow this section.)

Develop an idea of what you desire to achieve. Formu-

late it; criticize it; step back and look at it.   Wait a few
days, then look at it again--from various angles.   Vary your
components until you suddenly <u>know</u> that you have achieved
the unification you sought.   You will have created a tone
poem--a chromatic symphony--a work of art.

REFERENCES

Ancient Egyptian Materials and Industry, by A. Lucas. 4th
ed. London, 1962.

The Art of Arranging Flowers: A Complete Guide to Japanese Ikebana, by Shozo Sato. New York: Harry N.
Abrams, Inc., 1965. 366p. (oversize)

"Birch as a Source of Raw Material during the Stone Age,"
by E. Vogt, Proceedings of the Prehistory Society 15
(1949): 50-51. Also quoted in Plants and Archaeology,
by G. W. Dimbleby (see below).

The Book of Shrubs, by William Carey Grimm. New York:
Bonanza Books, 1957. 522p.

Bouquets That Last, by Emily Brown. New York: Hearth-
side Press, Inc., 1970. 176p.

The Complete Book of Dried Arrangements, by Raye Miller
Underwood. New York: Bonanza Books, 1952. 193p.

Decorating with Plant Crafts and Natural Materials, by Phyllis Pautz. Garden City, N.Y.: Doubleday, 1971.
239p.

Decorating with Pods and Cones, by Eleanor Van Rensselaer.
Princeton, N.J.: D. Van Nostrand Co., Inc., 1957.
179p.

Decorating with Seed Mosaics, Chipped Glass, and Plant Materials, by Eleanor Van Rensselaer. Princeton, N.J.:
D. Van Nostrand Co., Inc., 1960. 214p.

Design with Plant Material, by Marian Aaronson. London:
Grower Books, 1972. 119p.

Dried Flower Arrangement, by Edwin Rohrer. Princeton:
N.J.: Van Nostrand Reinhold, 1973. 89 p.

Dried Flower Arrangements from Garden, Bush and Seashore, by Nancy Millard. 6th ed. New York: Crescent Books, 1973. 128p.

Dried Flowers from Antiquity to the Present: A History and Practical Guide to Flower Drying, by Leonard Karel. Metuchen, N. J.: The Scarecrow Press, Inc., 1973. 184p.

Dried Flowers: The Art of Preserving and Arranging, by Nina de Yarburgh-Bateson. New York: Charles Scribner's Sons, 1972. 169p.

Flowers, Space and Motion, by Helen Van Pelt Wilson. New York: Simon and Schuster, 1971. 128p.

Forever Flowers, by Rejean Metzler. New York: Charles Scribner's Sons, 1972. 157p.

The Garden of Gourds, by L. H. Bailey. New York: The Macmillan Co., 1937. 134p.

Getting Started in Dried Flower Craft, by Barbara H. Amlick. New York: Bruce Publishing Co., 1971. 70p.

"Grasses, Ornamental, " by George W. Oliver, Plant Culture (1900): 92-95.

"Grasses Ornamental, " in The Royal Horticultural Society, Dictionary of Gardening. 2d ed. Oxford: The Clarendon Press, 1956. Vol. II (CO-JA), p. 922.

"Grasses, The Pampas, " by Otto Stapf, Flora and Sylva III (1905): 171-175.

The Guide to Garden Shrubs and Trees (including Woody Vines), by Norman Taylor. Boston: Houghton Mifflin, 1965. 450p.

The Guide to Victorian Antiques, by Marguerite W. Yates. New York: Harper & Bros., 1949. 246p.

History of European Botanical Discoveries in China, by E. Bretschneider. 2 vols. London: Sampson Low, Marston & Co., Ltd., 1898. 1167p. (English reprint, 1935.)

History of Flower Arrangement, by Julia S. Berrall. Rev.
    ed. New York: Viking, 1968. 176p.

How to Decorate with Natural Materials, by Phyllis Pautz.
    Garden City, N.Y.: Doubleday, 1973. 239p. (Paper-
    back edition of Decorating with Plant Crafts...--see
    above.)

An Illustrated Manual of Pacific Coast Trees, by Howard E.
    McMinn and Evelyn Maino. Berkeley, Calif.: Univ.
    of California Pr., 1935. 409p.

The International Book of Trees, by Hugh Johnson. New
    York, London: Simon and Schuster, 1973. 288p. (over-
    size)

Keeping the Plants You Pick, by Laura Louise Foster. New
    York: Thomas Y. Crowell Co., 1970. 149p.

Making Gifts from Oddments and Outdoor Materials, by Betsey
    B. Creekmore. New York: Hearthside Press, Inc.,
    1970. 224p.

New Decorations with Pods, Cones and Leaves, by Eleanor
    Van Rensselaer. Princeton, N.J.: D. Van Nostrand
    Co., Inc., 1966. 199p.

North American Trees (exclusive of Mexico and Tropical
    United States), by Richard J. Preston, Jr. 2d ed.
    Ames, Iowa: Iowa State University Pr., 1961. 395p.

"On Grasses in Herbal Literature," by Agnes Arber, Darwini-
    ana 5 (1941), 20-30.

Plants and Archaeology, by G. W. Dimbleby. Pall Mall,
    London: John Baker, 1967. 187p.

[Pliny's] Natural History, with an English translation in Ten
    Volumes, by W. H. S. Jones. Cambridge: Mass.:
    Harvard University Pr., 1951. Book XXI, III and IV,
    p. 165; Book XXXV, XLV. 156, p. 375.

The Rockwell's New Complete Book of Flower Arrangement,
    by F. F. Rockwell and Esther C. Grayson. Garden
    City, N.Y.: Doubleday, 1960. 336p.

"Some Ornamental Plants of Macon County, Alabama," by

George W. Carver, Bulletin No. 16 (Experiment Station Tuskegee Normal and Industrial Institute, Tuskegee Institute, Alabama), 1909, p. 16.

Standard Cyclopedia of Horticulture, by L. H. Bailey. 3 vols. New York: Macmillan, 1929-1944. 3,639p.

Wonders of the Flora, by H. Acosta Kresken. Dayton, Ohio: Philip A. Kemper, 1879. 204p.

Wyman's Garden Encyclopedia, by Donald Wyman. New York: Macmillan, 1971. 1,222p.

Table 1

NATURALLY (i. e. Air-) DRIED GRASSES
AND GRAINS FOR ORNAMENTAL USES

Common Name(s)	Botanical Name	Comments
Agropyron Grasses	Agropyron species	
Alfa Grass Esparto Grass	Stipa tenacissima	
Animated Oat	Avena sterilis	
Annual Beard Grass	Polypogon monspeliensis	Densely bearded spikes
Arundinaria Grass	Arundinaria japonica	A bamboo. Very attractive
Bamboo	Arundinaria japonica	Very attractive
Bamboo	Bambusa palmata	Very attractive
Bamboo	Bambusa tesellata	Very attractive
Bamboo	Bambusa veitchii	Very attractive
Barley, Common	Hordeum vulgare	
Barley, Squirreltail Squirreltail grass	Hordeum jubatum	Bushy spikes
Barnyard Grass Cockspur Grass	Echinochloa crusgallii	One-sided flow- erclusters re- sembling spikes
Bear Grass Elk Grass	Xerophyllum tenax	
Beard Grass, Silver	Andropogon argenteus	Very attractive
Bearded Wheatgrass	Agropyron subsecundum	

35

Common Name(s)	Botanical Name	Comments

Bent Grass
  Marram Grass
  Mat Weed
	Ammophila arenaria	A beach grass

Bent Grass, Silky
  Corn Grass
  Wind-grass
	Panicum clandestinum	

Big Bluestem	Andropogon gerardii	
Big Quaking Grass Common " "	Briza maxima	Beautiful spike- lets

Black Grass
  Mousetail
  Slender Foxtail
	Alopecurus agrestis	

Blue Fescue	Festuca ovina glauca	
Blue Wild Rye	Elymus glaucus	Attractive
Bluebunch Wheatgrass	Agropyron spicatum	
Bluegrass, Kentucky	Poa pratensis	
Bluegrass, Sandberg	Poa secunda	
Bluestem, Big	Andropogon gerardii	
Bonnet Grass Red Top Grass	Agrostis alba	
Bottle Brush Grass	Hystrix species	
Bristly Foxtail Grass	Setaria verticillata	
Brome, California	Bromus carinatus	
Brome, Mountain	Bromus marginatus	
Brome, Rattle	Bromus brizaeformis	Very attractive
Brome Grass	Bromus macrostachys	Very attractive
Brome Grass	Bromus madritensis	Very attractive
Brome Grass, Corn	Bromus arvensis	Large spikelets
Buckwheat	Fagopyrum esculentum	
Bunchgrass	Andropogon scoparius	Tufts

Bunchgrass	Elymus condensatus	Tufts
Bunchgrass	Oryzopsis hymenoides	Tufts
Bunchgrass	Sporobolus aeroides	Tufts
Bunchgrass, Feather Needlegrass, Green	Stipa viridula	Tufts
California Brome	Bromus carinatus	
Canada Wild Rye	Elymus canadensis	
Canary Grass Reed Canary Grass	Phalaris arundinacea	Very attractive
Canary Grass, Reed Canary Grass	Phalaris arundinacea	
Cane Grass	Glyceria ramigera	Fluffy heads
Cinna	Cinna latifolia	Very attractive
Citronella Grass Ginger Grass	Cymbopogon nardus	Very attractive
Cladium	Cladium meyenii	Rust-colored
Cloud Grass	Agrostis nebulosa	Feathery panicle
Cock'sfoot Orchard Grass	Dactylis glomerata	
Cockspur Grass Barnyard Grass	Echinochloa crusagallii	One-sided flower clusters resembling spikes
Common Barley	Hordeum vulgare	
Common Cultivated Oat	Avena sativa	
Common Giant Fennel	Ferula communis	Very attractive
Common Reed Ditch Reed Giant Reed	Phragmites communis	Large bushy panicle
Common Wheat Two-grained wheat	Triticum vulgare	
Common Winter Cress	Barbarea vulgaris	
Corn Brome Grass	Bromus arvensis	Large spikelets

Common Name(s)	Botanical Name	Comments
Corn Grass Silky Bent Grass Wind-grass	Panicum clandestinum	
Corn, Indian Maize	Zea mays	
Cotton Grass Cotton Sedge	Eriophorum polystachion	Tufted spikes
Cotton Grass Hare's Tail Grass	Lagurus ovatus	Beautiful spike
Cotton Sedge Cotton Grass	Eriophorum polystachion	Tufted spikes
Couch Grass	Agropyron repens	
Cress, Common Winter	Barbarea vulgaris	
Crested Dog'stail Grass Dog'stail Grass	Cynosurus cristatus	
Crested Wheatgrass	Agropyron cristatum	
Crimson Fountain Grass	Pennisetum ruppelii	Very attractive
Crow-foot Grass	Dactyloctenum aegyptium	Attractive
Cynosurus [Grass]	Cynosurus elegans	Very attractive
Cyperus	Cyperus natalensis	Attractive
Dallis Grass	Paspalum dilatatum	Tall and tufted
Ditch Reed Common Reed Giant Reed	Phragmites communis	Large, bushy panicle
Dog'stail Grass Crested Dog'stail Grass	Cynosurus cristatus	
Dog'stail Grass, Crested Dog'stail Grass	Cynosurus cristatus	Attractive
Downy Oat Grass	Trisetum spicatum	
Dropseed	Sporobolus minutiflorus	A delicate panicle with minute spikelets

Dropseed, Sand	Sporobolus cryptandrus	Ample panicle
Dune Grass, European	Elymus arenarius	Tall and tufted
Elephant Grass Napier Grass	Pennisetum purpureum	
Elk Grass Bear Grass	Xerophyllum tenax	
Emmer Wheat	Triticum aestivum	
Esparto Grass Alfa Grass	Stipa tenacissima	
Eulalia Grass	Miscanthus sinensis	Very attractive
Eulalia Grass, Striped	Miscanthus sinensis variegatus	
European Dune Grass	Elymus arenarius	Tall and tufted
European Feather Grass Feather Grass	Stipa pennata	

False Oat Grass
  Tall Oat Grass
  Tall Meadow-Oat
	Arrhenatherum elatius	

Feather Bunchgrass Green Needlegrass	Stipa viridula	Tufts
Feather Grass	Stipa pennata	
Feather Grass, European	Stipa pennata	
Fennel, Common Giant	Ferula communis	Attractive
Fescue, Blue	Festuca ovina glauca	
Fescue, Meadow	Festuca elatior	
Fescue, Red	Festuca rubra	
Fescue, Sheep	Festuca ovina	
Fescue, Tall	Festuca elatior var. arundinacea	
Fescue, Various- Leaved	Festuca heterophylla	Attractive
Finger Grass	Chloris elegans	Attractive

Common Name(s)	Botanical Name	Comments
Finger Grass	Chloris polydactyla	Spikelets with long, silky hairs
Fountain Grass	Pennisetum setaceum	
Fountain Grass, Crimson	Pennisetum ruppelii	Very attractive
Foxtail Grass	Hordeum species	Spikes resemble brushes
Foxtail Grass	Lolium rigida	Spikes resemble brushes
Foxtail Grass, Bristly	Setaria verticillata	
Foxtail Grass, Meadow	Alopecurus pratensis	
Foxtail Millet	Setaria italica	
Gama Grass Sesame Grass	Tripsacum dactyloides	
German Panick-Grass	Panicum germanicum	

Giant Reed
  Common Reed
  Ditch Reed
	Phragmites communis	Large, bushy panicle

Giant Reed Grass	Arundo donax	Attractive
Giant Wild Rye	Elymus condensatus	
Ginger Grass Citronella Grass	Cymbopogon nardus	
Golden-Top Grass	Lamarckia aurea	Showy, one-sided panicles
Green Needlegrass Feather Bunchgrass	Stipa viridula	Tufts
Guinea Grass	Panicum maximum	
Gynerium	Gynerium saccharoides	Attractive
Hair Grass	Aira elegans	
Hair Grass, Tufted	Deschampsia caespitosa	Very decorative

Hair Grass, Wood	Deschampsia flexuosa	Very decorative
Hare's Tail Grass Cotton Grass	Lagurus ovatus	Beautiful
Herd's Grass (Timothy)	Phleum pratense	
Himalaya Fairy-Grass	Miscanthus nepalensis	Very attractive
Indian Corn (Maize)	Zea mays	
Indian Grass	Sorghastrum nutans	Beautiful
Indian Rice Water Rice	Zizania palustris	
Indian Rice-Grass	Oryzopsis hymenoides	
Intermediate Wheatgrass	Agropyron intermedium	
Italian Rye Grass	Lolium multiflorum	
Job's Tears Tear Grass	Coix-lacryma-jobi	Very attractive
Johnson Grass	Sorghum halepense	
Kentucky Bluegrass	Poa pratensis	
Lady's Thumb	Polygonum persicaria	
Lemon Grass	Cymbopogon citratus	Very attractive
Little Quaking Grass	Briza minima	Attractive
Love-Grass Teff	Eragrostis abyssinica	Very attractive
Love-Grass	Eragrostis suaveolens	Very attractive
Lyme Grass	Elymus species	
Maiden Grass	Miscanthus sinensis gracillinus	
Maize Indian Corn	Zea mays	

Marram Grass
    Bent Grass
    Mat Weed
	Ammophila arenaria	A beach grass

Mat Weed	Ammophila arenaria	A beach grass

Common Name(s)	Botanical Name	Comments
Bent Grass Marram Grass		
Meadow Fescue	Festuca elatior	
Meadow Foxtail Grass	Alopecurus pratensis	
Meadow Grass, Reed	Glyceria grandis	Attractive
Meadow-Oat, Tall Oat Grass, False Oat Grass, Tall	Arrhenatherum elatius	
Meadow Soft Grass Velvet Grass	Holcus lanatus	
Melic Grass	Melica ciliata	
Millet, Foxtail	Setaria italica	
Millet, Pearl	Pennisetum glaucum	Dense, cylindrical spikes
Millet, True	Panicum miliaceum	Very attractive
Millet Grass	Milium effusum	
Moor Grass	Molinia caerulea	
Mountain Brome	Bromus marginatus	
Mountain Rice	Oryzopsis racemosa	
Mousetail Blackgrass Slender Foxtail	Alopecurus agrestis	
Naked Oat	Avena nuda	
Napier Grass Elephant Grass	Pennisetum purpureum	
Natal Grass Ruby Grass	Tricholaena rosea	Very showy
Needle-and-Thread	Stipa comata	
Needlegrass, Green Bunchgrass, Feather	Stipa viridula	Tufts
Oat, Animated	Avena sterilis	

Common Name	Scientific Name	Type	Size	Description
Dwarf Pomegranate	Punica granatum	Pod		Orange, pointed
Dwarf St. Peter'swort	Hypericum suffruticosum	Capsule	3/16	
Dwarf Spiraea	Spiraea betulifolia	Pod		Small, usually in groups of 5, and persistent
Dwarf Spruce	Picea glauca conica	Cone		
Dwarf Sumac	Rhus copallina	Drupe	1/8 cluster	Round, red, hairy; large clusters; persistent
Dwarf Witch-Alder	Fothergilla gardenii	Capsule	Less than 1/2	Egg-shaped, woody, downy, 2-beaked
Dyer's Greenweed	Genista tinctoria	Pod		Globular to narrow oblong
Early Winter Cress	Barbarea verna	Pod		Longer than pod of winter cress
Earpod Tree (Elephant-ear Pods)	Enterolobium cyclocarpum	Pod		Round, fluted, black
East Indian Lotus	Nelumbo nucifera	Pod		Large
Eastern Arborvitae (American ") (Northern White Cedar)	Thuja occidentalis	Cone	1/3-1/2	
Eastern Coral-Bean	Erythrina herbacea	Pod	Up to 5	
Eastern Cottonwood (Poplar)	Populus deltoides	Capsule	1/3	In catkins 8"-12" long; ovoid

Common Name	Botanical Name	Type of Seed Vessel	Vessel Dimension (length x width) in inches	Comments
Eastern Hemlock (Canada Hemlock)	Tsuga canadensis	Cone	1/2-3/4	Ovoid
Eastern Hophornbeam	Ostrya virginiana	Capsule	1/4	Ovoid, strobiles 1"-1 1/2" long; flat
Eastern Larch (American Larch) (Tamarack)	Larix laricina	Cone	1/2-3/4	Oval or almost globular
Eastern Ninebark (Common Ninebark)	Physocarpus opulifolius	Pod		Reddish to brown; small, papery, inflated; 3 to 5 grouped in cluster
Eastern Red Cedar (Redcedar-Juniper)	Juniperus virginiana	Cone	1/4-1/3	Pea-size, bluish, berry-like
Eastern Redbud	Cercis canadensis	Pod	2-3	Smooth
Eastern Wahoo (Burning Bush)	Euonymus atropurpureus	Capsule	3/4 across	
Eastern White Pine	Pinus strobus	Cone	4-6	Cylindric, slender, often curved, reddish brown
Ebony	Pithecolobium flexicaule	Pod	4-6	
Egyptian Acacia (Gum Arabic Tree)	Acacia arabica	Pod		

Common name	Scientific name		Size	Description
Ekoa (False Koa)	Leucaena glauca	Pod	5-6	Round, fluted, black
Elephant-Ear Pods (Earpod Tree)	Enterolobium cyclocarpum	Pod		Nearly round; not hairy on the margin; clusters
Elm, English	Ulmus procera	Capsule	3/4	Conical, beaked, gray-ish-brown
Empress, China (Royal Paulownia)	Paulownia tomentosa	Capsule	1-1 1/2	Cylindric, light-brown
Engelmann Spruce	Picea engelmannii	Cone	1-3	Nearly round; not hairy on the margin; clusters
English Elm	Ulmus procera	Capsule	3/4	Rust
Enkianthus	Enkianthus campanulatus	Capsule	Small	Bristly; 10-ribbed
Epaulette-Tree	Pterostyrax hispida	Capsule	1/2	
Eriogonum	Eriogonum species	Pod		Dark red, with red caps
Eucalyptus	Eucalyptus erythrocorys	Pod		
Eucalyptus	Eucalyptus ficifolia	Pod	1/2-2	Almost globular
Eucalyptus (Red Flowering Gum)	Eucalyptus filifolia	Pod	1 1/2	
Eucalyptus	Eucalyptus forestiana	Pod		Ruddy

Common Name	Botanical Name	Type of Seed Vessel	Vessel Dimension (length x width) in inches	Comments
Eucalyptus	Eucalyptus macrocarpa	Pod		
Eucalyptus	Eucalyptus megacornuta	Pod		Red, green, tan caps
Eucalyptus	Eucalyptus preissiana	Pod		Four-fringed
Eucalyptus species	Eucalyptus species	Pod	Many sizes	Many shapes, colors; gray, green, brown
Euonymus	Euonymus yedoensis	Pod		4-lobed; pinkish; orange interior
European Beech	Fagus sylvatica	Bur	1/2	Nut enclosed in bur
European Glorybind	Convolvulus arvensis	Capsule		
European Larch	Larix decidua	Cone	1-1 1/2	
Evening Primrose, Common	Oenothera biennis	Pod		
Evergreen Candytuft	Iberis sempervirens	Pod		
Everlasting Pea	Lathyrus grandiflorus	Pod		
Evodia, Korean	Evodia daniellii	Pod		A collection of small, hooked pods

False Arborvitae	Thujopsis dolabrata	Cone	1/2	Egg-shaped
False Jessamine (Carolina Jessamine) (Yellow Jessamine)	Gelsemium sempervirens	Pod	1/2-3/4	Egg-shaped, flattened, short-beaked
False Koa (Ekoa)	Leucaena glauca	Pod	5-6	
False Saffron (Safflower)	Carthamus tinctorius	Pod		
False Wiliwili	Adenanthera pavonina	Pod		
False Willow	Baccharis angustifolia	Capsule		Clusters, small
Fern, Cinnamon	Osmunda cinnamomea	Capsule		Spike
Fern, Sensitive	Onoclea sensibilis	Capsule		Spike
Fetid Buckeye (Ohio Buckeye)	Aesculus glabra	Capsule	2	Often spiny
Fetterbush	Lyonia lucida	Capsule	3/16" wide	
Fetterbush, Mountain	Pieris floribunda	Capsule	3/16	Egg-shaped
Field Mustard (Wild ")	Brassica kaber pinnatifida	Pod	3/4	
Field Mustard (Rape)	Brassica rapa	Pod		

Common Name	Botanical Name	Type of Seed Vessel	Vessel Dimension (length x width) in inches	Comments
Field Pennycress	Thlaspi arvense	Pod		Clusters
Figwort	Scrophularia nodosa	Capsule		
Finger Pods ("Max Watson" Pods)	Eucalyptus lehmannii	Pod		
Fir, Alpine	Abies lasiocarpa	Cone	3-4	
Fir, Balsam	Abies balsamea	Cone	2-4	Oblong, cylindrical, purple
Fir, Bristlecone	Abies bracteata	Cone		
Fir, California Red	Abies magnifica	Cone	6-9	Oblong, cylindrical, purplish brown
Fir, Cascade	Abies amabilis	Cone	3 1/2-6	Oblong, dark purple
Fir, China	Cunninghamia lanceolata	Cone	2	Nearly round
Fir, Cilician	Abies cilicica	Cone	5-9	Stout, cylindrical, orange-brown
Fir, Common China	Cunninghamia lanceolata	Cone	1-2	Nearly round, brown, prickly, glossy
Fir, Douglas	Pseudotsuga taxifolia	Cone	2-4 1/2	Egg-shaped

Common Name	Scientific Name	Type	Size	Description
Fir, Fraser Balsam	Abies fraseri	Cone	1 1/2-2 1/2 x 1	Oblong-ovate or nearly oval, purple
Fir, Grand (Fir, Lowland White)	Abies grandis	Cone	2-4 1/2	Cylindrical, rounded at apex, greenish
Fir, Greek	Abies cephalonica	Cone	5-7	Cylindrical, slender, pointed, grayish brown
Fir, Hamalayan	Abies spectabilis	Cone	2-4	
Fir, Korean	Abies koreana	Cone		
Fir, Lowland White (Fir, Grand)	Abies grandis	Cone	2-4 1/2	Cylindrical, rounded at apex
Fir, Momi	Abies firma	Cone	3 1/2-5	
Fir, Nikko	Abies homolepis	Cone	2-4	
Fir, Noble	Abies procera	Cone	2-4-6	
Fir, Nordmann	Abies nordmanniana	Cone	4-6	Oblong, cylindrical or ellipsoidal, dark orange-brown or reddish-brown
Fir, Pacific Silver (Fir, Cascade)	Abies amabilis	Cone	3 1/2-6	
Fir, Red (Fir, California Red)	Abies magnifica	Cone	6-9	Oblong, cylindrical, purplish brown

Common Name	Botanical Name	Type of Seed Vessel	Vessel Dimension (length x width) in inches	Comments
Fir, Red	Abies nobilis	Cone	4-6	Oblong, cylindrical, purplish or olive-brown
Fir, Silver	Abies alba	Cone	3-5	
Fir, Spanish	Abies pinsapo	Cone	2-6	Cylindrical, slender, grayish brown, or brownish-purple
Fir, Subalpine	Abies lasiocarpa	Cone	2-4	Dark purple
Fir, Veitch	Abies veitchii	Cone	1 1/2-4	Cylindrical, slender, dark purple
Fir, White	Abies concolor	Cone	2-5	Oblong, gray-green, dark purple, or bright canary-yellow
Fireweed	Epilobium angustifolium	Capsule	2-3	
Fish-Poison Tree	Piscida erythrina	Pod	2-4	
Fiveleaf Akebia	Akebia quinata	Pod	2-5	Fleshy, purple
Flame Amherstia	Amherstia nobilis	Pod	6-8	
Flame Azalea	Azalea calendulaceum	Capsule	3/4	Egg-shaped, downy

Flame Bottle Tree	Brachychiton acerifolium	Pod		
Flamebean, Glory	Brownea grandiceps	Pod		Flat
Flame-Tree (Royal Poinciana)	Delonix regia	Pod	24 x 2	
Flame-Tree, Chinese	Koelreuteria formosana	Pod		Bladder-like, orange and red
Flannel Bush (Fremontia Flannelbush)	Fremontia californica	Capsule	1	Ovoid, densely woolly
Flax, Yellow	Linum virginianum	Pod		Small
Florida Anise Tree (Purple Anise)	Illicium floridanum	Capsule	1 1/4" across	Wheel-shaped cluster, greenish
Florida Chinkapin	Castanea alnifolia var. floridana	Bur	Less than 2	
Flower-of-an-Hour	Hibiscus trionum	Capsule		Hairy
Formosa Sweet Gum	Liquidambar formosana	Ball	1	Bristly
Fortune Keteleeria	Keteleeria fortunei	Cone	3-7	
Fothergilla, Large (Witch-Alder, Mountain)	Fothergilla major	Capsule	1/2	Downy, woody, egg-shaped; 2-beaked
Foxglove	Digitalis purpurea	Capsule		

Common Name	Botanical Name	Type of Seed Vessel	Vessel Dimension (length x width) in inches	Comments
Foxtail Pine	Pinus balfouriana	Cone	3-5	Dark purplish brown, pendulous, sub-cylindric
Fragrant Sumac	Rhus aromatica	Drupe	1 1/4 cluster	Persistent, round, red, densely hairy; clusters
Franklinia	Gordonia altamaha	Capsule	3/4	Woody; globe-shaped
Fraser Balsam Fir	Abies fraseri	Cone	1 1/2-2 1/2 x 1	Oblong-ovate or nearly oval; purple
Fraser Magnolia	Magnolia fraseri	"Pod"		
Fremont Cottonwood	Populus fremontii	Pod		Small; clusters
Fremontia Flannelbush (Flannelbush)	Fremontia californica	Capsule	1	Ovoid, densely woolly
French Honeysuckle	Hedysarum coronarium	Pod		
Fringed Gentian	Gentiana crinita	Capsule	3/8 x 1/8	
Fritillaria	Fritillaria meleagris	Pod		
Frostweed	Helianthemum canadense	Capsule		
Fuller's Teasel	Dipsacus fullonum	Capsule	2-3 x 1	Stiff, spiny fruit heads

Furze (Gorse)	Ulex europaeus	Pod	1/2	Brown; hairy
Gaillardia	Gaillardia amblyodon	Pod		
Galtonia (Summer Hyacinth)	Galtonia candicans	Pod		
Garlic-Mustard	Alliaria officinalis	Pod	1	4-sided
Gas Plant (Dittany)	Dictamnus albus	Pod		
Gazania (Treasure Flower)	Gazania rigens	Pod		
Gentian, Closed	Gentiana andrewsii	Capsule	3/8 x 1/8	
Gentian, Fringed	Gentiana crinita	Capsule	3/8 x 1/8	
Georgia Basil	Satureja georgiana	Capsule		Nutlets, small–in groups of 4 within calyx
Georgia Bush-Honey-suckle	Diervilla rivularis	Capsule	1/2	
Georgia Indigobush	Amorpha georgiana	Capsule	3/16	Smooth, or nearly so
Giant Arborvitae (Western Red Cedar)	Thuja plicata	Cone	1/2	Cluster
Giant Sequoia (Big Tree)	Sequoia gigantea	Cone	2-3 1/2	Egg-shaped

Common Name	Botanical Name	Type of Seed Vessel	Vessel Dimension (length x width) in inches	Comments
Giant Spider Plant	Cleome spinosa	Pod		Long, slender
Ginger, Wild	Asarum canadense	Pod		
Gladwin Iris	Iris foetidissima	Pod		
Globe Candytuft	Iberis umbellata	Pod		
Globe Flower, American	Trollius laxus	Pod		Small
Glory Flamebean	Brownea grandiceps	Pod		Flat
Glorybind, European	Convolvulus arvensis	Capsule		Bursts
Glory-of-the-Snow	Chionodoxa luciliae	Capsule		3-sided
Glossy Shower Senna	Cassia glauca	Pod		Flat, green, clusters
Godetia	Godetia amoena	Capsule		
Golden Chain Tree	Laburnum anagyroides	Pod	2	Thin, smooth
Golden Chinkapin	Castanopsis chrysophylla	Bur	1-1 1/2	Prickly spines; ovoid nut enclosed in a globose bur
Golden Clematis	Clematis tangutica	Capsule		Dry; showy; long, silky plume

Golden Heather	Hudsonia ericoides	Capsule		Small, enclosed by calyx
Golden Larch	Pseudolarix amabilis kaempferi	Cone	1-1 1/2	Flower-like
Golden Rain Tree	Koelreuteria paniculata	Pod	1 1/2-2	Yellowish green; papery; long clusters
Golden St. John'swort	Hypericum frondosum	Capsule	1/2-3/4	Egg-shaped
Golden-Shower Senna	Cassia fistula	Pod	12-48	Black
Gompholobium	Gompholobium polymorphum	Pod		Inflated; globular or nearly so
Gordonia (Loblolly Bay)	Gordonia lasianthus	Capsule	2/3	Silky, woody, oblong
Gorse (Furze)	Ulex europaeus	Pod	1/2	Brown; hairy
Grand Fir (Lowland White Fir)	Abies grandis	Cone	2-4 1/2	Cylindrical, rounded at apex, greenish
Grape Hyacinth	Muscari botryoides	Pod		
Grasswort, Starry	Cerastium arvense	Capsule		As long as, or longer than, the calyx
Gray Sage (Silver Sagebrush)	Artemisia cana	Capsule		

Common Name	Botanical Name	Type of Seed Vessel	Vessel Dimension (length x width) in inches	Comments
Great Plantain	Plantago major	Capsule		Spike
Great Rhododendron (Rosebay ")	Rhododendron maximum	Capsule	1/2	Reddish brown, sticky, woody, oblong-ovoid
Greek Fir	Abies cephalonica	Cone	5-7	Cylindrical, slender, pointed, grayish-brown
Green Alder	Alnus crispa	Cone-like		Nutlets
Green-Barked Acacia (Blue Paloverde)	Cercidium torreyanum	Pod	3-4 x 1/4-1/3	Oblong
Grevillea, Silk-Oak	Grevillea robusta	Pod		
Groundsel Tree, Sessile-Flowered	Baccharis glomeruliflora	Capsule		Small; clusters
Gum, Blue (Eucalyptus)	Eucalyptus globulus	Pod	1 1/2 x 1 1/2	
Gum, Formosa Sweet	Liquidambar formosana	Ball	1	Bristly
Gum, Red Flowering (Eucalyptus)	Eucalyptus filifolia	Pod	1 1/2 x 1 1/2	Almost globular
Gum, Redbox (Gum, Silver Dollar)	Eucalyptus polyanthemos	Pod	1/2	

Gum, Silver Dollar (Gum Redbox)	Eucalyptus polyanthemos	Pod	1/2	
Gum, Sweet	Liquidambar styraciflua	Ball-like	1-1 1/2 x 1-1 1/2	Bristly capsules; brown
Gum Arabic Tree (Egyptian Acacia)	Acacia arabica	Pod		
Gum Tree (Sandarach)	Callitris quadrivalvis	Cone	Less than 1	
Hairy Bush Honeysuckle	Diervilla rivularis	Capsule	1/4	Oblong
Hairy Laurel	Kalmia hirsuta	Capsule	1/8 diameter	
Hakea	Hakea saligna	Pod	1-1 1/4	
Hakea, Sweet	Hakea suaveolens	Pod		
Hala (Screw Pine)	Pandanus odoratissimus	Pod		
Handkerchief Tree (Davidia) (Dove Tree)	Davidia involucrata	Nut	1 1/4 x 1	Ovoid, ribbed, brown
Hardhack (Steeplebush)	Spiraea tomentosa	Pod		Small, usually in groups of 5, persistent
Hardy Catalpa (Northern Catalpa)	Catalpa speciosa	Pod	8-20	Cylindric, persistent, pencil-thick
Hazel Alder	Alnus rugosa	Cone-like	2/3	

Common Name	Botanical Name	Type of Seed Vessel	Vessel Dimension (length x width) in inches	Comments
Hazel Sterculia	Sterculia foetida	Pod	4	
Hazelnut, American	Corylus americana	Nut	3/4	Nut enclosed by a husk 1 1/2"-2 1/4" long
Hazelnut, Beaked	Corylus cornuta	Nut		Nut enclosed by a pair of bracts forming a bristly-hairy beak
Heath, Mountain	Phyllodoce caerulea	Capsule	3/16	Egg-shaped
Heather, Beach	Hudsonia tomentosa	Capsule		Small, enclosed by calyx
Heather, Golden	Hudsonia ericoides	Capsule		Small, enclosed by calyx
Heather, Mountain	Hudsonia montana	Capsule		Small, enclosed by calyx
Hedge Mustard	Sisymbrium officinale	Pod		Closely hugs stem
Hemlock, Canada (Hemlock, Eastern)	Tsuga canadensis	Cone	1/2-3/4	Ovoid
Hemlock, Carolina	Tsuga caroliniana	Cone	1-1 1/2	Oblong
Hemlock, Eastern (Hemlock, Canada)	Tsuga canadensis	Cone	1/2-3/4	Ovoid
Hemlock, Japanese	Tsuga diversifolia	Cone	1/2-1	Egg-shaped

Hemlock, Mountain	Tsuga mertensiana	Cone	2-3	Purple before maturity, brownish when ripe
Hemlock, Siebold	Tsuga sieboldii	Cone	1-1 1/2	Ovate
Hemlock, Western	Tsuga heterophylla	Cone	3/4-1	Oblong-ovoid
Hemp Sesbania	Sesbania macrocarpa	Pod	7-9	Flat
Hercules' Club	Aralia spinosa	Capsule	1/4	Showy clusters; ovoid, wrinkled, black berry-like
Hercules' Club	Zanthoxylum clava-herculis	Capsule		Reddish, small; black seeds
Hibiscus	Hibiscus syriacus	Capsule	5/16	Nearly globular, hairy, grayish-brown
Hibiscus, Chinese	Hibiscus rosa-sinensis	Capsule		
Hibiscus, Cotton Rose	Hibiscus mutabilis	Capsule		Rounded, hairy
Hickory, Bitternut	Carya cordiformis	Nut		Nut in pearshaped husk
Hickory, Oval Pignut (Hickory, Red)	Carya ovalis	Nut	1-1 1/4	Oval nut encased in a pear-shaped husk
Hickory, Pig Nut	Carya glabra	Nut		Nut encased in a pear-shaped husk
Hickory, Red (Hickory, Oval Pignut)	Carya ovalis	Nut	1-1 1/4	Oval nut encased in a pear-shaped husk

Common Name	Botanical Name	Type of Seed Vessel	Vessel Dimension (length x width) in inches	Comments
Himalayan Fir	Abies spectabilis	Cone	2-4	
Himalayan Pine	Pinus griffithii (excelsa)	Cone	6-12	Cylindric
Himalayan Spruce	Picea smithiana	Cone		
Hinoki White Cedar	Chamaecyparis obtusa	Cone	1/3-1/2	Rounded
Hoary Azalea	Azalea canescens	Capsule	3/4	
Holly, Desert	Atriplex hymenelytra	Pod		
Hollyhock	Althaea rosea	Capsule		
Honesty	Lunaria annua	Pod	2	Rounded, flat
Honey Locust	Gleditsia triancanthos	Pod		Persistent; sickle-shaped, twisted
Honey Mesquite	Prosopis glandulosa	Pod		
Honeysuckle, Cape	Tecomara capensis	Pod	2	
Honeysuckle, Hairy Bush	Diervilla rivularis	Capsule	1/4	Oblong
Honeysuckle, Northern Bush	Diervilla lonicera	Capsule	1/2	Egg-shaped, beaked

Common Name	Scientific Name	Type	Size	Shape/Description
Honeysuckle, Southern Bush	Diervilla sessilifolia	Capsule	5/8-3/4	Oblong
Hong Kong Orchid-Tree	Bauhinia blakeana	Pod	12	Flat
Hophornbeam, Eastern	Ostrya virginiana	Capsule	1/4	Ovoid, flat, unwinged nut. Spikes 1"-1 1/2" long
Hoptree (Wafer-Ash)	Ptelea trifoliata	Capsule	3/4	Papery, greenish; almost orbicular wing; drooping clusters up to 4 1/2" long; persistent
Horehound	Marrubium vulgare	Pod		
Horse-Chestnut	Aesculus carnea	Bur	1 1/2	
Horse-Chestnut, Red	Aesculus hippocastanum	Bur	1 1/2	
Horse-Radish Tree	Moringa pterygosperma	Pod	18	3-sided, light brown
Horsetail Beefwood	Casuarina equisetifolia	Cone	1/2 wide	Ellipsoidal
Hosta	Undulata picta	"Pod"		
Hutu	Barringtonia asiatica	Pod		
Hyacinth, Grape	Muscari botryoides	Pod		
Hyacinth Bean	Dolichos lablab	Pod	2 1/2	Flat
Hyssop	Hyssopus officinalis	Capsule		Spike

Common Name	Botanical Name	Type of Seed Vessel	Vessel Dimension (length x width) in inches	Comments
Iceland Poppy	Papaver nudicaule	Pod		
Incense Cedar (California Incense Cedar)	Libocedrus decurrens	Cone	3/4-1 1/2	
Indian Coral Bean (Tiger's Claw)	Erythrina variegata orientalis	Pod		Black
Indian Mallow	Abutilon theophrastii	Pod		
Indian Mustard (Chinese ")	Brassica juncea	Pod		
Indigo	Indigofera tinctoria	Pod	3/4-1	
Indigo, Blue Wild	Baptisia australis	Pod		
Indigo, Wild	Baptisia tinctoria	Pod		
Indigobush, Common	Amorpha fruticosa	Pod	3 1/2-4 1/2	Curved; large raised glands on fruit
Indigobush, Georgia	Amorpha georgiana	Pod	3/16	Smooth, or nearly so
Indigobush, Mountain	Amorpha glabra	Pod	5/16	Sparingly glandular-dotted
Indigobush, Schwerin	Amorpha schwerinii	Pod	3/16	Hairy

Common Name	Scientific Name	Type	Size	Description
Indigobush, Shining	Amorpha nitens	Pod		Curved, smooth
Iris	Iris siberica	Pod	2	
Iris, Bearded	Iris aphylla	Pod		
Iris, Blue Flag	Iris versicolor	Pod	1 1/2	
Iris, Gladwin	Iris foetidissima	Pod		
Iris, Spanish	Iris xiphium	Pod		
Iris, Yellow Flag	Iris pseudacorus	Pod		
Italian Cypress	Cupressus sempervirens	Cone	1-1 1/2	Oblong or nearly globose
Italian Stone Pine	Pinus pinea	Cone	5	Broadly ovate; chestnut-brown
Jacaranda	Jacaranda acutifolia	Pod		
Jacaranda, Sharpleaf	Jacaranda mimosifolia	Pod		
Jack Pine	Pinus banksiana	Cone	1-2	Conic-oblong, usually curved, pale yellow-brown and lustrous; Remains on tree 12-15 years
Japanese Black Pine	Pinus thunbergii	Cone	2-3	Conic-ovate, grayish brown
Japanese Hemlock	Tsuga diversifolia	Cone	1/2-1	Egg-shaped

Common Name	Botanical Name	Type of Seed Vessel	Vessel Dimension (length x width) in inches	Comments
Japanese Larch	Larix kaempferi	Cone	1	Yellowish, slender
Japanese Pagoda Tree	Sophora japonica	Pod	2-3	Conic-ovate to oblong, grayish brown
Japanese Red Pine	Pinus densiflora	Cone	2	Ovate or oblong-ovate, reddish brown
Japanese White Pine	Pinus parviflora	Cone	2-3	
Japanese Wisteria	Wisteria floribunda	Pod	4 1/2-7	Velvety, hanging
Jasmine	See Jessamine			
Jeffrey Pine	Pinus jeffreyi	Cone	6-12	Conic-ovate, light brown
Jelecote Pine	Pinus patula	Cone	2	
Jerusalem Sage	Phlomis fruticosa	Pod		Small
Jerusalem-Thorn	Parkinsonia aculeata	Pod	3-4	Approximately cylindrical but tapering at both ends
Jessamine, Carolina	Gelsemium sempervirens	Pod	1/2-3/4	Egg-shaped, flattened, short beaked
Jessamine, False	Gelsemium sempervirens	Pod	1/2-3/4	Egg-shaped, flattened, short-beaked

Jessamine, Scentless	Gelsemium rankinii	Pod	3/4 or more	Beaked
Jessamine, Yellow	Gelsemium sempervirens	Pod	1/2-3/4	Egg-shaped, flattened, short beaked
Jessamine, Yellow Star	Trachelospermum asiaticum	Pod		Slender, long
Jimson Weed	Datura stramonium	Pod	3/4 x 1/2	
Joshua Tree	Yucca brevifolia	Capsule		
Juniper, Alligator	Juniperus deppeana	Cone	1/3-1/2	Dry, berry-like. Fruit is sweet and edible.
Juniper, California	Juniperus californica	Cone	3/8-3/4	Dry, berry-like
Juniper, Common	Juniperus communis	Cone	Pea-size	Dry, berry-like, bluish-black
Juniper, Utah	Juniperus osteosperma	Cone	1/4-3/4	Dry, berry-like
Kaffir Bean	Schotia brachypetala	Pod		
Kakalaioa	Caesalpinia crista	Pod		Prickly, brown
Kale, Sea	Crambe maritima	Pod		
Kalm St. John'swort	Hypericum kalmianum	Capsule	3/4	Narrowly egg-shaped
Kapok Tree (Silk Cotton Tree)	Ceiba pentandra	Pod	3-6	Like a cucumber and leathery

Common Name	Botanical Name	Type of Seed Vessel	Vessel Dimension (length x width) in inches	Comments
Karo	Pittosporum crassifolium	Capsule	3/4-1 1/4	Woody or leathery; to-mentose
Katsura-Tree	Cercidiphyllum japonicum	Pod		Persistent; splitting
Kentucky Coffee Tree	Gymnocladus dioicus	Pod	8-12	Brownish; persistent
Kentucky Wisteria	Wisteria macrostachya	Pod	2 1/2-5	Bean-like, knobby
Keteleeria	Keteleeria davidiana	Pod	6-8	
Keteleeria, Fortune	Keteleeria fortunei	Cone	3-7	
Kidney Bean	Phaseolus vulgaris	Pod		
Knapweed	Centaurea dealbata	Capsule		
Knife Acacia	Acacia cultriformis	Pod	3	
Knobcone Pine	Pinus attenuata	Cone	3 1/2-6	Elongated-conical
Knotweed (Smartweed)	Polygonum custidatum	Capsule	1/8 x 1/16	Decorative, rose-colored calyx; winged fruit
Knotweed, Prostrate	Polygonum aviculare	Capsule		Clusters
Kobus Magnolia	Magnolia kobus	"Pod"	4-5	

Korean Evodia	Evodia daniellii	Pod		Small, hooked pods
Korean Fir	Abies koreana	Cone		
Korean Pine	Pinus koraiensis	Cone	4-6	Conic-oblong, yellowish brown
Koster Spruce	Picea pungens "Pendens"	Cone		
Koyama Spruce	Picea koyamaii	Cone		
Kudzu Vine	Pueraria hirsuta	Pod		Large
Labrador-Tea	Ledum groenlandicum	Capsule	1/2	
Laburnum, Scotch	Laburnum alpinum	Pod		Small, pea-like
Laburnum, Waterer	Laburnum x watereri	Pod		Pea-like
Lacebark Pine	Pinus bungeana	Cone	2-3	Conic-ovate, light yellowish brown
Lady's Thumb	Polygonum persicaria			Curved spikes; attractive.
Lagenaria	Lagenaria siceraria	Gourd		Hard-shelled, durable
Lambert Crazy-weed	Oxytropis lambertii	Pod		
Lamb's Ears	Stachys lanata	Capsule		Spikes
Larch, American (Tamarack) (Larch, Eastern)	Larix laricina	Cone	1/2-3/4	Oval or almost globular

Common Name	Botanical Name	Type of Seed Vessel	Vessel Dimension (length x width) in inches	Comments
Larch, Dunkeld	Larix eurolepis	Cone	1-1 1/2	
Larch, Eastern (Larch, American) (Tamarack)	Larix laricina	Cone	1/2-3/4	Oval or almost globular
Larch, European	Larix decidua	Cone	3/4-1 1/2	Oval, downy
Larch, Golden	Pseudolarix amabilis kaempferi	Cone	1-3 1/2	Egg-shaped, reddish brown
Larch, Japanese	Larix kaempferi	Cone	1	
Larch, Western	Larix occidentalis	Cone	1-1 1/2	Oblong
Large Fothergilla (Mountain Witch-Alder)	Fothergilla major	Capsule	1/2	Downy, woody, egg-shaped; 2-beaked
Larkspur	Delphinium ajacis	Capsule		Cluster; cup-shaped
Laurel, Hairy	Kalmia hirsuta	Capsule	1/8	Globose
Laurel, Mountain	Kalmia latifolia	Capsule	3/16	Globose, woody
Laurel, Pale	Kalmia polifolia	Capsule	1/8	Globose
Laurel, Sheep	Kalmia angustifolia	Capsule	1/8	Globose

Common Name	Scientific Name	Type	Size	Description
Leadplant	Amorpha canescens	Capsule	3/16	Densely hairy
Leadtree (White Popinac)	Leucaena glauca	Pod	6	Narrow, reddish
Leatherflower	Clematis viorna	Capsule	1 1/2	Brown plume-like tails
Leatherleaf	Chamaedaphne calyculata	Capsule	1/8	Roundish
Lebbeck Tree (Woman's Tongue Tree)	Albizzia lebbek	Pod	12	Flat; lustrous
Lemon	Citrus limon	Berry		Puncture ends; dry for several weeks; tint lemon-yellow
Lemon Bottlebrush	Callistemon lanceolatus	Capsule		Woody; very ornamental
Lentil	Lens culinaris	Pod	3/4	Flat
Lespedeza	Lespedeza species	Pod		
Leucadendron	Leucadendron plumosum	Pod		Reddish-brown
Leucadendron	Leucadendron sabulosum	Pod	2	Reddish-brown; ridged
Leucothoë, Coastal	Leucothoë axillaris	Capsule	3/16	Roundish
Leucothoë, Drooping	Leucothoë fontanesiana	Capsule	3/16	Roundish
Leucothoë, Recurved	Leucothoë recurva	Capsule	3/16	Roundish

Common Name	Botanical Name	Type of Seed Vessel	Vessel Dimension (length x width) in inches	Comments
Leucothoë, Swamp	Leucothoë racemosa	Capsule	3/16	Roundish
Leucothoës	Leucothoë species	Capsule	3/16	5-lobed, roundish
Licorice	Glycyrrhiza glabra	Pod		Flat
Lily, Day	Hemerocallis fulva	Pod		Can be dried at green and at tan stages
Lily, Peruvian	Alstroemeria pelegrina	Pod		
Lily, Plantain	Hosta species	Pod		
Lily, Yellow Turk's Cap	Lilium pyrenaicum	Pod		
Lily-of-the-Valley Tree (Lily Tree, White)	Crinodendron dependens	Pod		
Lily Tree, White (Lily-of-the-Valley Tree)	Crinodendron dependens	Pod		
Limber Pine	Pinus flexilis	Cone	3-10	Ovate to cylindric-ovate; light-brown
Lipstick	Bixa arborea	Pod	1/2 x 2	Rounded sections
Lipstick	Bixa orellana	Pod	2	Prickly, pointed, reddish-brown

Littleleaf Pea-Tree	Caragana microphylla	Pod		
Loblolly Bay (Gordonia)	Gordonia lasianthus	Capsule	2/3	Woody, silky, oblong
Loblolly Pine	Pinus taeda	Cone	3-5	Conic-oblong, light reddish brown
Locust, Black	Robinia pseudoacacia	Pod	3-4	Pea-like, smooth
Locust, Bristly	Robinia hispida	Pod	1 1/2-2 1/2	
Locust, Clammy	Robinia viscosa	Pod	2-3	Sticky
Locust, Common Honey	Gleditsia triacanthus	Pod	12-18	Persistent; sickle-shaped; twisted
Lodgepole Pine	Pinus contorta latifolia	Cone	3/4-3	
Longleaf Pine	Pinus palustris	Cone	6-10	Cylindric, dull brown
Looking-Glass Tree	Heritiera littoralis	Pod		
Loosestrife, Swamp	Decodon verticillatus	Capsule	1/4	Urn-shaped
Lotus, American	Nelumbo lutea	Pod	2-3 at top	Large, brown, round at top, fluted; conical
Lotus, East Indian	Nelumbo nucifera	Pod		Large
Love-in-a-Mist	Nigella damascena	Pod		

Common Name	Botanical Name	Type of Seed Vessel	Vessel Dimension (length x width) in inches	Comments
Lowland White Fir (Grand Fir)	Abies grandis	Cone	2-4 1/2	Cylindrical, rounded at apex, greenish
Luffa	Luffa acutangula	Gourd		Sharply-angled, hard-shelled, durable
Luffa	Luffa cylindrica	Gourd		Cylindrical, hard-shelled, durable
Lupine	Lupinus species	Pod		Pea-like
Lyonia, Rusty	Lyonia ferruginea	Capsule	1/4"	Egg-shaped
Lycopodium	Lycopodium cernuum	Pod		
Macedonian Pine (Balkan Pine)	Pinus peuce	Cone	3 1/2-8	Cylindric
Macnab Cypress	Cupressus macnabiana	Cone	3/4-1	Oblong
Macrozamia	Macrozamia reidlei	Pod		Rich orange
Magnolia	Magnolia loebneri 'Merrill'	"Pod"	2	
Magnolia	Magnolia rostrata	"Pod"	5-6 x 1 1/2	Fruits borne in cone-like clusters; the individual fruits are pod-like

Magnolia	Magnolia sargentiana robusta	"Pod"	7-8	
Magnolia, Anise	Magnolia salicifolia	"Pod"		
Magnolia, Bigleaf	Magnolia macrophylla	"Pod"		
Magnolia, "Campbell"	Magnolia campbellii mollicomata	"Pod"		
Magnolia, Dawson	Magnolia dawsoniana	"Pod"		
Magnolia, Fraser	Magnolia fraseri	"Pod"		
Magnolia, Kobus	Magnolia kobus	"Pod"	4-5	
Magnolia, Oyama	Magnolia sieboldii	"Pod"	1 1/2	Crimson
Magnolia, Purple Lily	Magnolia liliflora 'nigra'	"Pod"		Brown, oblong
Magnolia, Saucer	Magnolia x soulangeana	"Pod"		Cucumber-like
Magnolia, Shinyleaf	Magnolia nitida	"Pod"		
Magnolia, Southern	Magnolia grandiflora	"Pod"	4	Rusty-woolly
Magnolia, Sprenger	Magnolia sprengeri	"Pod"		
Magnolia, Star	Magnolia stellata	"Pod"	2	Red
Magnolia, Swampbay (" Sweetbay)	Magnolia virginiana	"Pod"	2-5	Fruits borne in cone-like clusters; the individual fruits are pod-like, red

Common Name	Botanical Name	Type of Seed Vessel	Vessel Dimension (length x width) in inches	Comments
Magnolia, Sweetbay (" Swampbay)	Magnolia virginiana	"Pod"	2-5	Fruits borne in cone-like clusters; the individual fruits are pod-like, red
Magnolia, Umbrella	Magnolia tripetala	"Pod"		
Magnolia, Veitch	Magnolia x veitchii	"Pod"		
Magnolia, Watson	Magnolia x watsonii	"Pod"		
Magnolia, Whiteleaf Japanese	Magnolia obovata	"Pod"		
Magnolia, Wilson	Magnolia wilsonii	"Pod"		
Magnolia, Yulan	Magnolia denudata	"Pod"	3-4	Brownish, cylinder-like
Mahogany, West Indies	Swietenia mahogani	Capsule	3-4 x 3-4	Woody
Mahogany Tree	Swietenia mahogani	Capsule	3-4 x 3-4	Pear-shaped
Maleberry	Lyonia ligustrina	Capsule	1/8	Roundish
Mallow, Indian	Abutilon theophrastii	Pod		
Mallow, Rose	Hibiscus moscheutos	Capsule		
Maltese Cross	Lychnis chalcedonica	Capsule		

Common Name	Scientific Name	Type	Size	Description
Mango	Mangifera indica	Seed; firm skin		
Maples	Acer species	Samara		
Martynia	Craniolaria annua	Capsule		
Matilija Poppy	Romneya coulteri	Pod		
Maunaloa	Canavalia microcarpa	Pod		Thick, 3-ridged
"Max Watson" Pods (Finger Pods)	Eucalyptus lehmannii	Pod		
Maximowicz Pea-Tree	Caragana maximowicziana	Pod		
Mayten	Maytenus boaria	Capsule		Leathery
Meadow Saffron (Autumn Crocus)	Colchicum autumnale	Pod		
Meadowsweet, Broadleaf	Spiraea latifolia	Pod		Small, usually in groups of 5; persistent
Meadowsweet, Narrow-Leaf	Spiraea alba	Pod		Small, usually in groups of 5; persistent
Medicago	Medicago scutellata	Pod		Unusual shapes - spirally twisted
Melaleuca, Cajeput	Melaleuca leucadendron	Capsule		
Menziesia, Allegheny	Menziesia pilosa	Capsule	3/16	Bristly-hairy, egg-shaped

Common Name	Botanical Name	Type of Seed Vessel	Vessel Dimension (length x width) in inches	Comments
Mescal-Bean	Sophora secundiflora	Pod	1-7 x 1/2	Oblong, tapered at the ends, woody; dense, hoary, matted hairs
Mesquite	Prosopis juliflora	Pod	4-9 x 1/4-1/2	Linear, flat, leathery
Mesquite, Honey	Prosopis glandulosa	Pod		
Metasequoia	Metasequoia glyptostroboides	Cone	3/4	Oval
Mexican Buckeye	Ungnadia speciosa	Capsule	2 wide	Leathery, reddish brown
Mexican Cypress	Cupressus lusitanica	Cone	1/2 wide	Globular
Mexican Handtree	Chiranthodendron pentadactylon	Pod		
Mexican Pinyon Pine (Pinyon Pine)	Pinus cembroides	Cone	1 1/2-2	Sub-globose, lustrous brown
Mexican Prickle-Poppy	Argemone mexicana	Pod		Prickly
Michelia	Michelia compressa	Cone-like	2	Spike; leathery
Milk Vetch	Astragalus alopecuroides	Pod		Unusual shapes like snails and caterpillars

Milkweed, Butterfly	Asclepias tuberosa	Pod	3-4	
Milkweed, Common	Asclepias syriaca	Pod	3-4	
Milkweed, Swamp	Asclepias incarnata	Pod	3-4	
Millet	Panicum miliaceum	Drupe-like		Sumac-like clusters
Mimosa (Silk Tree)	Albizzia julibrissin	Pod	6	
Mock-Cucumber, Wild	Echinocystis lobata	Pod	2	Papery, puffed
Modoc Cypress	Cupressus bakeri	Cone	Less than 1	Gray
Moerheim Spruce	Picea "Moerheimii"	Cone		
Momi Fir	Abies firma	Cone	3 1/2-5	
Monkeypod Tree (Rain Tree)	Samanea saman	Pod	8	Reddish brown or black, fat, nubby; narrow, woody ridge
Monkey-Puzzle	Araucaria araucana	Cone	6-8 wide	Egg-shaped
Montbretia	Tritonia crocosmaeflora	Pod		
Monterey Cypress	Cupressus macrocarpa	Cone	1-1 1/2	Globular or oblong
Monterey Pine	Pinus radiata	Cone	3-5 1/2	Conic-ovate
Moonflower	Calonyction aculeatum	Pod		

Common Name	Botanical Name	Type of Seed Vessel	Vessel Dimension (length x width) in inches	Comments
Moreton Bay Chestnut	Castanospermum australe	Pod	2 x 9	
Morning Glory, Ceylon	Ipomoea tuberosa	Pod		
Mound-Lily Yucca	Yucca gloriosa	Capsule	2-3	
Mountain Azalea (Roseshell azalea)	Azalea roseum	Capsule	3/4	
Mountain Camellia	Stewartia ovata	Capsule	5/8	Egg-shaped, pointed, sharply 5-angled
Mountain Fetterbush	Pieris floribunda	Capsule	3/16	Egg-shaped
Mountain Heath	Phyllodoce caerulea	Capsule	3/16	Egg-shaped
Mountain Heather	Hudsonia montana	Capsule		Small, enclosed by calyx
Mountain Hemlock	Tsuga mertensiana	Cone	2-3	Cylindric-oblong, violet-purple before maturity, brownish when ripe
Mountain-Immortelle	Erythrina poeppigiana	Pod	5	

Mountain Indigobush	Amorpha glabra	Pod	5/16	Glandular-dotted
Mountain Laurel	Kalmia latifolia	Capsule	3/16	Roundish
Mountain Lover	Pachystima canbyi	Capsule	1/8	Small
Mountain Ninebark	Physocarpus monogynus	Pod		
Mountain St. John'swort	Hypericum buckleyi	Capsule	1/4-3/8	Egg-shaped, pointed
Mountain Silverbell	Halesia monticola	Drupe	2	Winged
Mountain Sweet Pepper-bush	Clethra acuminata	Capsule	3/16	Slightly egg-shaped or roundish
Mountain Witch-Alder (Large Fothergilla)	Fothergilla major	Capsule	1/2	Downy, woody, egg-shaped; 2-beaked
Mourning Cypress	Chamaecyparis funebris	Cone	1/2-1/2	Globular
Mullein, Common	Verbascum thapsus	Capsule		Spikes - may be over 12" long and durable; with small capsules
Mungo-Bean	Phaseolus mungo	Pod	2	
Mustard, Black	Brassica nigra	Pod	1	Hugs stem
Mustard, Chinese (" Indian)	Brassica juncea	Pod		
Mustard, Field (" Wild)	Brassica kaber pinnatifida	Pod	3/4	

Common Name	Botanical Name	Type of Seed Vessel	Vessel Dimension (length x width) in inches	Comments
Mustard, Indian (" Chinese)	Brassica juncea	Pod		
Mustard, Field (Rape)	Brassica rapa	Pod		
Mustard, Hedge	Sisymbrium officinale	Pod		Closely hugs stem
Mustard, Tower	Arabis glabra	Pod		Long, bushy, and hugs stem
Mustard, White	Brassica hirta	Pod		Bristling and ends in a flattened beak
Mustard, Wild (" Field)	Brassica kaber pinnatifida	Pod	3/4	
Myrtle-Leaf St. John'swort	Hypericum myrtifolium	Capsule	1/4	Egg-shaped
Naked-Flowered St. John'swort	Hypericum nudiflorum	Capsule	1/4	Narrowly egg-shaped
Nandin	Nandina domestica	Pod		Whitish, in clusters
Narrow-leaf Meadow-sweet	Spiraea alba	Pod		Small, usually in groups of 5 and persistent

Nasturtium, Vermilion	Tropaeolum speciosum	Capsule		Dull red
Naudin	Gomphocarpus textilis	Pod		Large, bladder-like
Neillia species	Neillia species	Capsule		Small
New Jersey Tea	Ceanothus americanus	Pod	Less than 1/2	3-lobed, cup-shaped; persistent
New Zealand Sophora	Sophora tetraptera	Pod	7	4-winged
Night Flower	Nyctanthes arbor-tristis	Pod	3/4	
Nightshade, Bitter	Solanum dulcamara	Capsule	2 x 2	Round globe
Nikko Fir	Abies homolepis	Cone	2-4	
Ninebark, Common (" Eastern)	Physocarpus opulifolius	Pod		Small, reddish to brown, inflated, papery, 3 to 5 on each stalk in cluster
Ninebark, Eastern (" Common)	Physocarpus opulifolius	Pod		Small, reddish to brown, inflated, papery, 3 to 5 on each stalk in cluster
Ninebark, Mountain	Physocarpus monogynus	Pod		Small
Nipplewort	Lapsana communis	Pod		
Noble Fir	Abies procera	Cone	2-6	
Nordmann Fir	Abies nordmanniana	Cone	4-6	Oblong, cylindrical or

Common Name	Botanical Name	Type of Seed Vessel	Vessel Dimension (length x width) in inches	Comments
Nordmann Fir (cont.)				ellipsoidal, dark orange-brown or reddish-brown
Norfolk Island Pine	Araucaria excelsa	Cone		Woody
Northern Bush-Honey-suckle (Dwarf Bush-Honeysuckle)	Diervilla lonicera	Capsule	1/2	Oblong, egg-shaped, slender
Northern Catalpa (Hardy Catalpa)	Catalpa speciosa	Pod	8-20	Cylindric, persistent, pencil-thick
Northern White Cedar (Eastern Arborvitae) (American ")	Thuja occidentalis	Cone	1/3-1/2	
Norway Pine (Red Pine)	Pinus resinosa	Cone	1 1/2-2 1/2	Conic-ovate, light brown
Norway Spruce	Picea abies	Cone	5-7	Smooth, cylindric-oblong, light brown
Nutmeg, California (Torreya, ")	Torreya californica	Drupe-like	1-1 1/2	Single seed; green-to-purple envelope
Oak, Tanbark (Tanoak)	Lithocarpus densiflorus	Acorn		Enclosed in a shallow,

densely hairy and matted, spiny, cup-like sheath lined with a lustrous red pubescence

Common Name	Scientific Name	Fruit	No.	Description
Oak	Quercus species	Acorn	1/2-1	Round to elliptical; enclosed in a cup
Octopus Tree	Brassaia actinophylla	Pod		
Ohio Buckeye (Fetid buckeye)	Aesculus glabra	Capsule	2	Usually spiny
Okra	Hibiscus esculentus	Pod	12	Long, cylindrical
Oleander	Nerium oleander	Pod		
Orange	Citrus aurantium	Berry		Puncture ends, dry for several weeks, tint orange
Orange Trumpet	Pyrostegia venusta	Pod	12	
Orchid-Tree	Bauhinia variegata	Pod	12	
Orchid-Tree, Hong Kong	Bauhinia blakeana	Pod	12	Flat
Oriental Arborvitae	Thuja orientalis	Capsule		Cluster of small, horned capsules
Oriental Bittersweet	Celastrus orbiculatus	Berry		Orange-red

Common Name	Botanical Name	Type of Seed Vessel	Vessel Dimension (length x width) in inches	Comments
Oriental Poppy	Papaver orientale	Pod	3/8 x 3/16	
Oriental Spruce	Picea orientalis	Cone	2-3 1/2	Cylindric-ovate, brownish-violet
Osage Orange	Maclura pomifera	Drupe-like	4-5	Numerous small drupes grown together into a multiple fruit resembling an orange; becomes woody
Oswego Tea (Bee Balm) (Wild Bergamot)	Monarda fistulosa	Capsule		
Oval Pignut Hickory (Red Hickory)	Carya ovalis	Nut		Encased in an oval husk 1"-1 1/4" in diameter
Oxeye, Sea	Borrichia frutescens	Capsule		Small
Oyama Magnolia	Magnolia sieboldii	"Pod"	1 1/2	Crimson
Ozark Chinkapin	Castanea ozarkensis	Bur	1-1 1/2	
Pacific Dogwood	Cornus nuttallii	Pod		
Pacific Silver Fir (Cascade Fir)	Abies amabilis	Cone	3 1/2-6	Oblong, dark purple

Pagoda Tree, Japanese	Sophora japonica	Pod	2-3	Yellowish; slender
Painted Buckeye	Aesculus sylvatica	Pod	1-1 1/2 wide	
Palay Rubbervine	Cryptostegia grandiflora	Pod		Borne in pairs, angled, pointed
Pale Laurel	Kalmia polifolia	Capsule	1/8 wide	
Pale Sweet-Shrub	Calycanthus fertilis	Pod	1 x 3/4	
Palm	Cocos nucifera	Nut		Coconut
Palm, Royal	Oreodoxa regia	Nut		Small coconut
Paloverde, Blue	Cercidium torreyanum	Pod	3-4 x 1/4-1/3	Oblong
Paper Birch	Betula papyrifera	Cone-like	1-1 1/2	Spikes on slender peduncles; nutlets
Paradise Flower (Catclaw Acacia)	Acacia greggii	Pod	2-6 x 1/2-3/4	Linear, oblong, flat, much curved and contorted
Paradise Poinciana	Caesalpinia gilliensii	Pod		
Parasol Tree, Chinese	Firmiana simplex	Pod	3-5	Leaf-like
Parrot's Bill	Clianthus puniceus	Pod		
Parry Pinyon Pine	Pinus quadrifolia	Cone	1-2 1/2	Sub-globose, chestnut-brown, lustrous

Common Name	Botanical Name	Type of Seed Vessel	Vessel Dimension (length x width) in inches	Comments
Paulownia, Royal (Empress, China)	Paulownia tomentosa	Capsule	1-1 1/2	Conical, beaked, grayish brown
Paw Paw, Common	Asimina triloba	Pod	2-5	Brown, oblong
Paw Paw, Dwarf	Asimina parviflora	Pod	1-2	Brown, oblong
Pea, Common Shamrock	Parochetus communis	Pod	1	
Pea, Everlasting	Lathyrus grandiflorus	Pod		
Pea, Perennial Sweet (Pea, Everlasting Sweet)	Lathyrus latifolius	Pod		
Pea, Pigeon (Cajan)	Cajanus cajan	Pod	3	
Pea, Sweet	Lathyrus odoratus	Pod		
Peach	Prunus persica	Pit		Football-shaped
Peachleaf Willow	Salix amygdaloides	Capsule	1/4	Globose-conic, 2"-long clusters
Pearl Acacia	Acacia podalyriae	Pod		
Pea-Tree, Littleleaf	Caragana microphylla	Pod		

Common Name	Scientific Name	Fruit	Size	Shape/Arrangement
Pea-Tree, Maximowicz	Caragana maximowicziana	Pod		
Pea-Tree, Pygmy	Caragana pygmaea	Pod	3/4-1 1/4	
Pea-Tree, Siberian	Caragana arborescens	Pod	1 1/2-2	Stalked
Pecan	Carya olivaeformis	Nut	1 1/2-2 1/2	Cylindrical
Peltophorum	Peltophorum inerme	Pod	3	
Pennycress, Field	Thlaspi arvense	Pod		Clusters
Pennycress, Roundleaf	Thlaspi rotundifolium	Pod		Clusters
Pentstemon (Beard-Tongue)	Penstemon fruticosus	Pod		
Peony, Garden	Paeonia officinalis	Capsule		Flower-like
Peony, Tree	Paeonia suffruticosa	Pod		
Pepper	Capsicum frutescens	Pod		
Pepperbush, Mountain Sweet	Clethra acuminata	Capsule	3/16	Roundish or slightly egg-shaped, on nodding stalks
Pepperbush, Sweet	Clethra alnifolia	Capsule	1/8	Roundish, on ascending stalks
Pepperbush, Woolly Sweet	Clethra alnifolia var. tomentosa	Capsule		Roundish, on ascending stalks

Common Name	Botanical Name	Type of Seed Vessel	Vessel Dimension (length x width) in inches	Comments
Pepper-Grass	Lepidium virginicum	Pod		Clusters
Peppermint Tree (Willow Myrtle)	Agonis flexuosa	Capsule		Woody
Perennial Sweet Pea (Everlasting Sweet Pea)	Lathyrus latifolius	Pod		
Peru-Balsam Tree	Myroxylon peruiferum	Pod		
Peruvian Lily	Alstroemeria pelegrina	Pod		
Phlomis	Phlomis viscosa	Capsule		
Pigeon Pea	Cajanus cajan	Pod	3	
Pignut Hickory	Carya glabra	Nut		Pear-shaped husks
Pinckneya	Pinckneya pubens	Capsule	3/4	Roundish
Pincushion Flower (Caucasian Scabious)	Scabiosa caucasica	Pod		
Pine, Aleppo	Pinus halepensis	Cone	2 1/2-3 1/2	Conic-ovate or conic-oblong, yellowish brown
Pine, Apache	Pinus engelmannii	Cone	2 1/2-6	

		Cone-like Pod		
Pine, Australian	Casuarina cunninghamiana			
Pine, Austrian	Pinus nigra	Cone	2-3 1/2	Ovate, yellowish brown, glossy
Pine, Balkan (Pine, Macedonian)	Pinus peuce	Cone	3 1/2-8	Cylindric
Pine, Big Cone (Pine, Coulter)	Pinus coulteri	Cone	9-14	Cylindric-ovate, yellowish brown
Pine, Bishop	Pinus muricata	Cone	2-4 1/2	Oblong-ovate, oblique at base, chestnut-brown
Pine, Bristlecone	Pinus aristata	Cone	1 3/4-3 3/4	Cylindric-ovate
Pine, Canary	Pinus canariensis	Cone	4-9	Cylindric-ovate
Pine, Chihuahua	Pinus leiophylla var. chihuahuana	Cone	1 1/2-2	Ovoid
Pine, Cluster	Pinus pinaster	Cone	3 3/4-7 3/4	Broadly-ovate, greenish yellow, lustrous
Pine, Colorado Pinyon	Pinus edulis	Cone	1-2 1/2	
Pine, Coulter (Pine, Big Cone)	Pinus coulteri	Cone	9-14	Cylindric-ovate, yellowish brown
Pine, Digger	Pinus sabiniana	Cone	6-10	Oblong-ovate, light red-brown

Common Name	Botanical Name	Type of Seed Vessel	Vessel Dimension (length x width) in inches	Comments
Pine, Eastern White	Pinus strobus	Cone	4-6	Cylindric, slender, often curved, reddish-brown
Pine, Foxtail	Pinus balfouriana	Cone	3-5	Dark purplish-brown, pendulous, sub-cylindric
Pine, Himalayan	Pinus griffithii (excelsa)	Cone	6-12	Cylindric
Pine, Italian Stone	Pinus pinea	Cone	5	Broadly ovate, chestnut-brown
Pine, Jack	Pinus banksiana	Cone	1-2	Conic-oblong, usually curved, pale yellow-brown and lustrous. Remains on the tree 12-15 years
Pine, Japanese Black	Pinus thunbergii	Cone	2-3	Conic-ovate, grayish brown
Pine, Japanese Red	Pinus densiflora	Cone	2	Conic-ovate to oblong, grayish brown
Pine, Japanese White	Pinus parviflora	Cone	2-3	Ovate or oblong-ovate, reddish brown
Pine, Jeffrey	Pinus jeffreyi	Cone	6-12	Conic-ovate, light brown

Common Name	Scientific Name	Type	Size	Description
Pine, Jelecote	Pinus patula	Cone	2	Elongated-conical
Pine, Knobcone	Pinus attenuata	Cone	3 1/2-6	Conic-oblong, yellowish brown
Pine, Korean	Pinus koraiensis	Cone	4-6	Conic-ovate, light yellowish brown
Pine, Lacebark	Pinus bungeana	Cone	2-3	Ovate to cylindric-ovate, light brown
Pine, Limber	Pinus flexilis	Cone	3-10	Conic-oblong, light reddish brown
Pine, Loblolly	Pinus taeda	Cone	3-5	
Pine, Lodgepole	Pinus contorta latifolia	Cone	3/4-3	Cylindric, dull-brown
Pine, Longleaf	Pinus palustris	Cone	6-10	Cylindric
Pine, Macedonian (Pine, Balkan)	Pinus peuce	Cone	3 1/2-8	
Pine, Mexican Pinyon (Pine, Pinyon)	Pinus cembroides	Cone	1 1/2-2	Sub-globose, lustrous brown
Pine, Monterey	Pinus radiata	Cone	3-5 1/2	Conic-ovate
Pine, Norfolk Island	Araucaria excelsa	Cone		Woody
Pine, Norway (Pine, Red)	Pinus resinosa	Cone	1 1/2-2 1/2	Conic-ovate, light brown

Common Name	Botanical Name	Type of Seed Vessel	Vessel Dimensions (length x width) in inches	Comments
Pine, Parry Pinyon	Pinus quadrifolia	Cone	1-2 1/2	Sub-globose, chestnut-brown, lustrous
Pine, Pinyon (Pine, Mexican Pinyon)	Pinus cembroides	Cone	1 1/2-2	Sub-globose, lustrous brown
Pine, Pitch	Pinus rigida	Cone	1 1/2-4	Ovate, light brown
Pine, Pond	Pinus serotina	Cone	2-2 1/2	Sub-globose to ovate, rounded or pointed at apex, light yellow-brown
Pine, Ponderosa (Pine, Western Yellow)	Pinus ponderosa	Cone	2 1/2-6	Ovate-oblong, light reddish or yellowish brown, lustrous
Pine, Red (Pine, Norway)	Pinus resinosa	Cone	1 1/2-2 1/2	Conic-ovate, light brown
Pine, Sand	Pinus clausa	Cone	2-3 1/2	Conic-ovate, often oblique at base, dark reddish brown. Closed for 3-4 years after ripening
Pine, Scotch	Pinus sylvestris	Cone	1 1/2-2 1/2	Conic-oblong, grayish or reddish brown
Pine, Scrub	Pinus virginiana	Cone	1 1/2-2 1/2	Conic-oblong, reddish

(Pine, Virginia)				brown
Pine, Shore	Pinus contorta	Cone	1-3	Ovate or conic-ovate, very oblique at base, light yellow-brown, lustrous
Pine, Shortleaf	Pinus echinata	Cone	1 1/2-2 1/2	
Pine, Singleleaf Pinyon	Pinus monophylla	Cone	1-2 1/2	Broadly ovate, greenish yellow, lustrous
Pine, Slash	Pinus caribaea	Cone	4-5	
Pine, Spruce	Pinus echinata	Cone	Less than 1 1/2-2 1/2	Conic-oblong, dull brown
Pine, Sugar	Pinus lambertiana	Cone	10-20	Cylindric, often slightly curved, light brown, lustrous
Pine, Swiss Mountain	Pinus mugo	Cone	3/4-2 1/4	Ovate or conic-ovate
Pine, Swiss Stone	Pinus cembra	Cone	2 1/2-3 1/2	Ovate, obtuse, light brown
Pine, Table Mountain	Pinus pungens	Cone	2 1/2-3 1/2	Conic-ovate, oblique at base, light brown
Pine, Torrey	Pinus torreyana	Cone	4-6	Broadly-ovate, chocolate-brown
Pine, Umbrella	Sciadopitys verticillata	Cone	3-5	Woody

Common Name	Botanical Name	Type of Seed Vessel	Vessel Dimension (length x width) in inches	Comments
Pine, Virginia (Pine, Scrub)	Pinus virginiana	Cone	1 1/2-2 1/2	Conic-oblong, reddish brown
Pine, Washoe	Pinus washoensis	Cone	9-12	Egg-shaped
Pine, Western White	Pinus monticola	Cone	4-11	Cylindric, slender, slightly curved, yellowish brown
Pine, Western, Yellow (Pine, Ponderosa)	Pinus ponderosa	Cone	2 1/2-6	Ovate-oblong, light reddish or yellowish brown, lustrous
Pine, White	Pinus strobus	Cone	4-7	Cylindric, slender, often curved, reddish-brown
Pine, Whitebark	Pinus albicaulis	Cone	4-6	
Pink Azalea (Pinxter Flower)	Azalea nudiflorum	Capsule	1/2-3/4	Narrowly egg-shaped; hairy
Pink Flame Bottle Tree	Brachychiton discolor	Pod		
Pinkshell Azalea	Azalea vaseyi	Capsule	1/2	Elliptical
Pinxter Flower (Pink Azalea)	Azalea nudiflorum	Capsule	1/2-3/4	Narrowly egg-shaped; hairy
Pinyon Pine	Pinus cembroides	Cone	1 1/2-2	Sub-globose, lustrous

(Mexican Pinyon Pine)				brown
Pipe, Dutchman's	Aristolochia durior	Capsule	5/8-3	Cylindrical, 6-ribbed
Pipevine, Woolly	Aristolochia tomentosa	Capsule	2-3	Ribbed, more or less woolly
Pipsissewa	Chimaphila umbellata	Capsule	3/16	
Pitch Pine	Pinus rigida	Cone	1 1/2-4	Ovate, light brown
Pittosporum species	Pittosporum species	Capsule	1/2-1 1/4 wide	Clusters, yellow, heart-shaped
Plane Tree, American (Buttonwood) (Sycamore)	Platanus occidentalis	"Ball"	1 x 1	
Plantain, Great	Plantago major	Capsule		Spike
Plantain Lily	Hosta species	Pod		
Plum-Yew	Cephalotaxus drupacea	Drupe-like	1	Oval; green
Poinciana, Paradise	Caesalpinia gilliensii	Pod		Flat
Poinciana, Royal (Flame Tree)	Delonix regia	Pod		Flat
Pokeweed	Phytolacca americana	Calyx		
Pomegranate, Dwarf	Punica granatum	Pod		Orange, pointed

Common Name	Botanical Name	Type of Seed Vessel	Vessel Dimension (length x width) in inches	Comments
Pond Pine	Pinus serotina	Cone	2-2 1/2	Sub-globose to ovate, rounded or pointed at apex, light yellow-brown
Ponderosa Pine (Western Yellow Pine)	Pinus ponderosa	Cone	2 1/2-6	Ovate-oblong, light reddish or yellowish brown, and lustrous
Poplar (Eastern Cottonwood)	Populus deltoides	Capsule	1/3	Ovoid; catkins 8"-12" long
Poplar (Quaking Aspen)	Populus tremuloides	Capsule	1/4	Curved, narrowly conical, fruiting branches about 3" long
Poplar, Balsam	Populus tacamahaca	Capsule	1/4	Ovoid to oblong
Poplar, Tulip (Poplar, Yellow) (Tulip Tree)	Liriodendron tulipifera	Pod	2 x 3/4	Persistent
Poplar, Yellow (Poplar, Tulip) (Tulip Tree)	Liriodendron tulipifera	Pod	2 x 3/4	Persistent
Poplar, White	Populus alba	Pod		Small
Poppy, California	Eschscholtzia californica	Pod	3-4	

Poppy, Iceland	Papaver nudicaule	Pod		
Poppy, Matilija	Romneya coulteri	Pod		
Poppy, Oriental	Papaver orientale	Pod	3/8 x 3/16	
Poppy, Prickly	Argemone grandiflora	Pod		
Port Oxford Cedar	Chamaecyparis lawsoniana	Cone	1/3-1/2	
Powderpuff-Redhead	Calliandra haematocephala	Pod		Flat, curled
Prickly Poppy	Argemone grandiflora	Pod		
Primrose, Common Evening	Oenothera biennis	Pod		
Primrose	Primula species	Capsule		
Prince's Feather	Polygonum orientale	Pod		Decorative stem and calyx
Prince's Plume, Desert	Stanleya pinnata	Pod		
Prostrate Knotweed	Polygonum aviculare	Capsule		Clusters
Prostrate Sandmyrtle	Leiophyllum buxifolium var. prostratum	Capsule	1/16	Egg-shaped
Protea	Protea arborea	Pod		Gold-colored
Protea	Protea mellifera	Pod		Reddish-brown

Common Name	Botanical Name	Type of Seed Vessel	Vessel Dimension (length x width) in inches	Comments
Purple Anise (Florida Anise Tree)	Illicium floridanum	Capsule	1 1/4 wide	Wheel-shaped cluster, greenish
Purple Cone Spruce	Picea purpurea	Cone		
Purple Lily Magnolia	Magnolia liliflor 'nigra'	"Pod"		Brown, oblong
Purple Sage (" Sagebrush)	Artemisia tridentata	Pod		
Purple Sagebrush (Purple Sage)	Artemisia tridentata	Pod		
Pussywillow	Salix discolor	Pod		
Pygmy Pea-Tree	Caragana pygmaea	Pod	3/4-1 1/4	
Quaking Aspen (Poplar)	Populus tremuloides	Capsule	1/4	Curved, narrowly conical; fruiting branches about 3" long
Queen Crape-Myrtle	Lagerstroemia speciosa	Pod		Cluster
Queen Palm	Arecastrum romanzoffi-anum	Pod		Boat-shaped
Radish	Raphanus sativus	Pod		Dark, fat

Radish, Wild	Raphanus raphanistrum	Pod		Conspicuously beaded
Rain Tree, Golden	Koelreuteria paniculata	Pod	1 1/2-2	Yellowish-green; papery; long clusters
Raintree Saman	Samanea saman	Pod	6-8	
Rangoon Creeper	Quisqualis indica	Capsule	1	Leathery, 5-angled
Rape (Field Mustard)	Brassica rapa	Pod		
Rattlebox	Crotalaria sagittalis	Pod		Black. Seeds rattle in the pod
Recurved Leucothoë	Leucothoë recurva	Capsule		3/16 diameter
Red Alder	Alnus rubra	Cone-like		Nutlets. Spikes 1/2"-1 1/4" long, woody, persistent, ovoid to oblong; orange peduncles
Red Basil	Satureja coccinea	Capsule		Nutlets - small - groups of 4 in calyx
Red Bauhinia	Bauhinia punctata	Pod	5	
Red Buckeye	Aesculus pavia	Pod	1-2	Egg-shaped; leathery, smooth
Red Cedar, Western (Giant arborvitae)	Thuja plicata	Cone	1/2	Clusters

Common Name	Botanical Name	Type of Seed Vessel	Vessel Dimension (length x width) in inches	Comments
Red Cedar-Juniper (Eastern Red Cedar)	Juniperus virginiana	Cone	1/4-1/3	Pea-size, bluish, berry-like
Red Fir	Abies nobilis	Cone	4-6	Oblong, cylindrical, purplish or olive-brown
Red Fir (California Red Fir)	Abies magnifica	Cone	6-9	Oblong, cylindrical, purplish brown
Red Flowering Gum (Eucalyptus)	Eucalyptus filifolia	Pod	1 1/2 x 1 1/2	Almost globular
Red Hickory	Carya ovalis	Nut	1-1 1/4	Oval nut encased in a pear-shaped husk
Red Horse-Chestnut	Aesculus hippocastanum	Bur	1 1/2	
Red Pine (Norway Pine)	Pinus resinosa	Cone	1 1/2-2 1/2	Conic-ovate, light brown
Red Spruce	Picea rubra	Cone	1 1/4-2	Oblong, green when young, light reddish-brown and glossy when mature
Redbox Gum (Silver Dollar Gum)	Eucalyptus polyanthemos	Pod	1/2	
Redbud, American	Cercis canadensis	Pod	3	Flat, narrow

Redbud, Eastern	Cercis canadensis	Pod	2-3	Flat
Redroot	Ceanothus ovatus	Capsule		Small, 3-lobed, cup-shaped, persistent
Redroot, Small-leaf	Ceanothus microphyllus	Capsule		Small, 3-lobed
Redwood, California	Sequoia sempervirens	Cone	1	
Redwood, Dawn	Metasequoia glyptostroboides	Cone	1	
Reed, Bur	Sparganium species	Bur-like		Globose
Rhododendron, Great (" , Rosebay)	Rhododendron maximum	Capsule	1/2	Sticky, reddish-brown, woody, oblong-ovoid
Rhododendron, Rosebay (" , Great)	Rhododendron maximum	Capsule	1/2	Sticky, reddish-brown, woody, oblong-ovoid
Rhodora	Rhododendron canadense	Capsule	1/2-5/8	Oblong
River Birch	Betula nigra	Cone-like		Nutlets. Spikes (Strobiles)
Robinia	Robinia "Idaho"	Pod		
Rose, Wood	Ipomoea tuberosa	Pod-like		Brown, globular, with dried and hardened sepals resembling a rose
Rose Acacia	Robinia hispida	Pod	2-3	Bristly

Common Name	Botanical Name	Type of Seed Vessel	Vessel Dimension (length x width) in inches	Comments
Rose Mallow	Hibiscus moscheutos	Capsule		
Rose	Rosa species	Hip	1/2-3/4	Nearly-round to oval; red; dark-red; scarlet, brown; black
Rosebay Rhododendron (Great Rhododendron)	Rhododendron maximum	Capsule	1/2	Sticky, reddish-brown, woody, oblong-ovoid
Rosemary, Bog	Andromeda glaucophylla	Capsule	3/16	Turban-shaped
Rose-of-Sharon	Hibiscus syriacus	Capsule	1 1/2 x 1/2	5-valved
Roseshell Azalea (Mountain Azalea)	Azalea roseum	Pod	3/4	Oblong, downy, hairy
Roundleaf Pennycress	Thlaspi rotundifolium	Pod		Clusters
Round-Podded St. John'swort	Hypericum cistifolium	Capsule	3/16	Egg-shaped
Royal Paulownia (China Empress)	Paulownia tomentosa	Capsule	1-1 1/2	Conical, beaked, grayish-brown
Royal Poinciana (Flame Tree)	Delonix regia	Pod	24 x 2	
Rubbervine, Palay	Cryptostegia grandiflora	Pod		Borne in pairs, pointed, angled

Common Name	Scientific Name	Type	Size	Description
Rue, Common	Ruta graveolens	Capsule		4 to 5-lobed
Running Strawberry Bush	Euonymus obovatus	Capsule		Rough-warty, pale orange-red
Safflower (False Saffron)	Carthamus tinctorius	Pod		
Saffron, Meadow (Crocus, Autumn)	Colchicum autumnale	Pod		
Sage	Salvia officinalis	Pod		
Sage, Gray (Sagebrush, Silver)	Artemisia cana	Capsule		
Sage, Jerusalem	Phlomis fruticosa	Pod		Small
Sage, Purple (Sagebrush, Purple)	Artemisia tridentata	Pod		
Sagebrush, Purple (Sage, Purple)	Artemisia tridentata	Pod		
Sagebrush, Silver (Sage, Gray)	Artemisia cana	Capsule		
Sageretia	Sageretia minutiflora	Capsule	1/4 wide	Roundish, dark purple, with 3 leathery nutlets
St. Andrew's Cross	Hypericum hypericoides	Capsule	1/4	

Common Name	Botanical Name	Type of Seed Vessel	Vessel Dimension (length x width) in inches	Comments
St. Andrew's Cross	Hypericum strogalum	Capsule	1/4	Conical, egg-shaped, pointed
St. John'swort, Bed-straw	Hypericum galioides	Capsule	1/4	Narrowly egg-shaped
St. John'swort, Bushy	Hypericum densiflorum	Capsule	Less than 1/4	Egg-shaped
St. John'swort, Golden	Hypericum frondosum	Capsule	1/2-3/4	Narrowly egg-shaped
St. John'swort, Kalm	Hypericum kalmianum	Capsule	3/8	Egg-shaped, pointed
St. John'swort, Mountain	Hypericum buckleyi	Capsule	1/4-3/8	Egg-shaped
St. John'swort, Myrtle-Leaf	Hypericum myrtifolium	Capsule	1/4	Narrowly egg-shaped
St. John'swort, Naked-flowered	Hypericum nudiflorum	Capsule	1/4	Egg-shaped
St. John'swort, Round-podded	Hypericum cistifolium	Capsule	3/16	Egg-shaped
St. John'swort, Shrubby	Hypericum spathulatum	Capsule	1/2-3/4	Egg-shaped, pointed
St. John'swort, Straggling	Hypericum dolabriforme	Capsule	1/4	

St. John'swort, Tutsan	Hypericum androsaemum	Capsule		Red at one stage
St. Peter'swort	Hypericum stans	Capsule	3/8	Egg-shaped
St. Peter'swort, Dwarf	Hypericum suffruticosum	Capsule	3/16	
Sakhalin Spruce	Picea glehnii	Cone	2-3	
Salt Tree	Halimodendron halodendron	Pod		
Salvia	Salvia species	Pod		
Saman, Raintree	Samanea saman	Pod	6-8	
Sand Pine	Pinus clausa	Cone	2-3 1/2	Conic-ovate, often oblique at base, dark reddish brown. Closed for 3-4 years after ripening.
Sandalwood	Santalum album	Pod		
Sandarach (Gum Tree)	Callitris quadrivalvis	Cone	Less than 1	
Sandbox Tree	Hura crepitans	Pod	3	Ridged, rounded like a pumpkin
Sandweed	Hypericum fasciculatum	Capsule	1/4	Narrowly egg-shaped
Sandweed	Hypericum lloydii	Capsule	1/4	
Sandweed	Hypericum nitidum	Capsule	1/4	

Common Name	Botanical Name	Type of Seed Vessel	Vessel Dimension (length x width) in inches	Comments
Sandweed	Hypericum reductum	Capsule	1/4	
Santolina	Santolina chamaecyparissus	Pod		
Sassafrass	Sassafrass albidum	Berry-like		Bluish-black, with red stalks
Saucer Magnolia	Magnolia x soulangeana	"Pod"		Cucumber-like
Sausage-Tree	Kigelia pinnata	Pod	Up to 24 x 5	Sausage-shaped
Sawara White Cedar	Chamaecyparis pisifera	Cone	1/4	
Saxegothea	Saxegothea conspicua	Cone		
Scarlet Clematis	Clematis texensis	Capsule		Plumed
Scarlet-Runner Bean	Phaseolus coccineus	Pod	12	
Scentless Yellow Jessamine	Gelsemium rankinii	Pod	3/4 or more	Beaked
Schotia	Schotia latifolia	Pod	1 1/2-2 1/2	Oblong or broad linear
Schrenk Spruce	Picea schrenkiana	Cone	3-4	Cylindric-ovate
Schwerin Indigobush	Amorpha schwerinii	Pod	3/16	Hairy

Scorpiuris	Scorpiuris species	Pod		Unusual shapes; worm-like
Scotch Broom	Cytisus scoparius	Pod	1-2	Slender; usually hairy
Scotch Laburnum	Laburnum alpinum	Pod		Small, pea-like
Scotch Pine	Pinus sylvestris	Cone	1 1/2-2 1/2	Conic-oblong, grayish or reddish brown
Screw Bean (Tornillo)	Prosopis pubescens	Pod	1	Twisted like a corkscrew, thin, glossy, light tan
Screw Pine (Hala)	Pandanus odoratissimus	Pod		
Scrub Pine (Virginia Pine)	Pinus virginiana	Cone	1 1/2-2 1/2	Conic-oblong, reddish brown
Scurf-Pea	Psoralea species	Pod		
Sea Kale	Crambe maritima	Pod		
Sea Oxeye	Borrichia frutescens	Capsule		Small
Sea Poppy	Glaucium flavum	Capsule		Long, narrow
Sea-Rocket	Cakile edentula	Pod		Distinctive
Seaside Alder	Alnus maritima	Cone-like		Nutlets borne in the axils of woody scales

Common Name	Botanical Name	Type of Seed Vessel	Vessel Dimension (length x width) in inches	Comments
Sea-Urchin Tree	Hakea laurina	Pod	1 1/4	
Sebastian Bush	Sebastiana ligustrina	Capsule	1/4	3-lobed, yellowish-green
Seedbox	Ludwigia alternifolia	Pod		Short, squarish
Senna, Bladder	Colutea arborescens	Pod		Inflated, flat, papery
Senna, Glossy Shower	Cassia glauca	Pod		Flat, green clusters
Senna, Golden Shower	Cassia fistula	Pod	12-48	Black
Senna, Wild	Cassia marilandica	Pod	4	
Sensitive Fern	Onoclea sensibilis	Capsule		Spike
Sequoia, Giant (Bigtree)	Sequoia gigantea	Cone	2-3 1/2	Egg-shaped or nearly oval
Serbian Spruce	Picea omorika	Cone	2-2 1/2	Glossy-brown
Sesbania, Hemp	Sesbania macrocarpa	Pod	7-9	Flat
Sessile-Flowered Groundsel Tree	Baccharis glomeruliflora	Capsule		Small, clusters
Shamrock-Pea, Common	Parochetus communis	Pod	1	

Sharpleaf Jacaranda	Jacaranda mimosifolia	Pod		
Sheep Laurel	Kalmia angustifolia	Capsule	1/8 diameter	
Shell Flower (Bells-of-Ireland)	Molucella laevis	Capsule		
She-Oak	Casuarina stricta	Cone	1/2	Elliptical
Shepherd's Purse	Capsella rubella	Pod		Triangular, flattened
Shepherd's Purse, Alien	Capsella bursa-pastoris	Pod		Triangular, rounded
Shining Indigobush	Amorpha nitens	Pod		Curved, smooth
Shinyleaf Magnolia	Magnolia nitida	"Pod"		
Shinyleaf Yellowhorn	Xanthoceras sorbifolium	Bur		Like that of horse-chest-nut
Shore Pine	Pinus contorta	Cone	1-3	Ovate or conic-ovate, very oblique at base, light yellow-brown, lustrous
Shortleaf Pine	Pinus echinata	Cone	1 1/2-2 1/2	
Shower-Tree, Coral	Cassia grandis	Pod	24	Cylindrical, dark-brown
Showy Crotalaria	Crotalaria spectabilis	Pod	2	"Rattle-box," rounded
Shrub Althaea	Hibiscus syriacus	Capsule	5/16	Grayish brown

Common Name	Botanical Name	Type of Seed Vessel	Vessel Dimension (length x width) in inches	Comments
Shrub Yellow Root	Xanthorhiza simplicissima	Capsule		Small, light yellow, inflated 1-seeded, 4-8 grouped together
Shrubby St. John'swort	Hypericum spathulatum	Capsule	1/2-3/4	Egg-shaped
Siberian Iris	Iris siberica	Pod	1 1/4-1 1/2	Brown
Siberian Pea-Tree	Caragana arborescens	Pod	1 1/2-2	Stalked
Siberian Spruce	Picea obovata	Cone	2 1/2	Oblong-ovate, light brown
Sicklepod	Cassia tora	Pod		Long, sickle-shaped
Siebold Hemlock	Tsuga sieboldii	Cone	1-1 1/2	Ovate
Silk Tree (Mimosa)	Albizzia julibrissin	Pod	5-6 x 1	Flat
Silk Vine	Periplaca graeca	Pod	3-5	Narrow, smooth
Silk-Cotton Tree (Kapok Tree)	Ceiba pentandra	Pod	3-6	Like a cucumber, leathery
Silk-Oak Grevillea	Grevillea robusta	Pod		
Silky Camellia	Stewartia malachodendron	Capsule	1/2	Roundish
Silver Dollar Gum (Redbox Gum)	Eucalyptus polyanthemos	Pod	1/2	

Common Name	Scientific Name	Type	Size	Description
Silver Fir	Abies alba	Cone	3-5	
Silver Sagebrush (Gray Sage)	Artemisia cana	Capsule		
Silver Spruce	Picea pungens "Argentea"	Cone		
Silverbell, Carolina	Halesia caroliniana	Drupe	1-2	Winged; somewhat oblong
Silverbell, Mountain	Halesia monticola	Drupe	2	Winged
Singleleaf Pinyon Pine	Pinus monophylla	Cone	1-2 1/2	Broadly ovate, greenish yellow, lustrous
Sitka Spruce	Picea sitchensis	Cone	2-4	
Slash Pine	Pinus caribaea	Cone	4-5	
Small-Leaf Redroot	Ceanothus microphyllus	Capsule		Small, 3-lobed
Smoketree, American	Cotinus obovatus	Pod-like		Small, lopsided
Smooth Sumac	Rhus glabra	Drupe	1 1/8 cluster	Persistent; round; compact clusters; red, sticky hairy
Smooth Strawberry Bush	Euonymus americanus	Capsule	1/2-3/4 across	Rough-warty, crimson
Snapdragon	Antirrhinum majus	Pod		Cluster

Common Name	Botanical Name	Type of Seed Vessel	Vessel Dimension (length x width) in inches	Comments
Sneezeweed	Helenium autumnale	Capsule		
Soapbark Tree	Quillaja saponaria	Pod		Green
Soaptree Yucca (Spanish Bayonet)	Yucca elata	Capsule	1 1/2-2	Thin and woody, erect, oblong, light brown
Sogabark Peltophorum	Peltophorum pterocarpum	Pod	2-3	Reddish-brown
Sophora, New Zealand	Sophora tetraptera	Pod	7	4-winged
Sophora, Vetch	Sophora davidii	Pod		
Sorrel Tree (Sourwood Tree)	Oxydendrum arboreum	Capsule	1/3-1/2	Persistent; hairy
Sourwood Tree (Sorrel Tree)	Oxydendrum arboreum	Capsule	1/3-1/2	Persistent; hairy
Southern Bush-Honey-suckle	Diervilla sessilifolia	Capsule	5/8-3/4	Oblong
Southern Catalpa	Catalpa bignonoides	Pod	9-15	Cylindric, pencil-thick
Southern Magnolia	Magnolia grandiflora	"Pod"	4	Rusty-woolly
Southern Plume	Elliottia racemosa	Capsule	3/8	Roundish

Common Name	Scientific Name	Fruit Type	Size	Description
Southern White Cedar	Chamaecyparis thyoides	Cone	1/4–1/2	Globose
Soy–Bean	Glycine max	Pod	2–3 x 1/2	
Spanish Bayonet	Yucca aloifolia	Capsule	2 1/2–4	
Spanish Bayonet (Soaptree Yucca)	Yucca elata	Capsule	1 1/2–2	Thin, woody, oblong, erect, light brown
Spanish Bluebell	Scilla hispanica	Pod		
Spanish Broom	Genista hispanica	Pod	1	Hairy, oblong-like
Spanish Broom	Spartium junceum	Pod	2–4	Flattened; hairy
Spanish Chestnut	Castanea sativa	Bur	2–4	Prickly; contains 1–3 nuts; roundish
Spanish Dagger	Yucca gloriosa	Capsule	2–3	
Spanish Fir	Abies pinsapo	Cone	2–6	Cylindrical, slender, grayish brown or brownish-purple
Spanish Iris	Iris xiphium	Pod		
Speckled Alder	Alnus incana	Cone-like	2/3	Winged nutlets borne in the axils of woody scales
Spicebush	Calycanthus occidentalis	Pod		

Common Name	Botanical Name	Type of Seed Vessel	Vessel Dimension (length x width) in inches	Comments
Spindle Tree, Aldenham	Euonymus europaeus 'Aldenhamensis'	Pod	3/4	4-lobed; red or pink; bright orange interior; persistent
Spindle Tree, Winged	Euonymus alatus	Pod		Usually in 4's; showy; bright scarlet interior; persistent
Spiraea	Spiraea species	Pod		Small, usually in groups of 5; persistent
Spiraea, Dwarf	Spiraea betulifolia	Pod		Small, usually in groups of 5; persistent
Spiraea, Virginia	Spiraea virginiana	Pod		Small, usually in groups of 5; persistent
Sponge-Tree	Acacia farnesiana	Pod		
Spotted Cowbane	Cicuta maculata	Pod		
Spotted Wintergreen	Chimaphila maculata	Capsule	1/4	Round
Spreading Dogbane	Apocynum androsaemifolium	Pod	3-8	Narrow; in pairs
Sprenger Magnolia	Magnolia sprengeri	"Pod"		
Spruce, Alberta	Picea albertiana	Cone		

Spruce, Alcock	Picea bicolor	Cone	2 1/4-4 3/4	
Spruce, Aldenham	Picea aldenhamensis	Cone		
Spruce, Baker	Picea pungens 'Bakeri'	Cone		
Spruce, Big Cone	Pseudotsuga macrocarpa	Cone		
Spruce, Black	Picea mariana	Cone	1/2-1 1/2	Oval-oblong, globose-ovate, dark purple when young, dull grayish brown when mature
Spruce, Black Hills	Picea glauca densata	Cone		
Spruce, Blue Colorado	Picea pungens glauca	Cone	2 1/4-4 1/2	
Spruce, Brewer	Picea breweriana	Cone	2 1/2-5	
Spruce, Colorado	Picea pungens	Cone		
Spruce, Dragon	Picea asperata	Cone	3 1/4-5	Cylindric-oblong, fawn-gray changing to chest-nut-brown
Spruce, Dwarf	Picea glauca conica	Cone		
Spruce, Engelmann	Picea engelmannii	Cone	1-3	Cylindric, light brown
Spruce, Himalayan	Picea smithiana	Cone	5-7	Dark brown, glossy
Spruce, Koster	Picea pungens "Pendens"	Cone		

Common Name	Botanical Name	Type of Seed Vessel	Vessel Dimension (length x width) in inches	Comments
Spruce, Koyama	Picea koyamaii	Cone		
Spruce, Moerheim	Picea 'Moerheimii'	Cone		
Spruce, Norway	Picea abies	Cone	5-7	Cylindric-oblong, light brown, smooth
Spruce, Oriental	Picea orientalis	Cone	2-3 1/2	Cylindric-ovate, brownish-violet
Spruce, Purple Cone	Picea purpurea	Cone		
Spruce, Red	Picea rubra	Cone	1 1/4-2	Oblong, green when young, light reddish brown and glossy when mature
Spruce, Sakhalin	Picea glehnii	Cone	2-3	
Spruce, Schrenk	Picea schrenkiana	Cone	3-4	Cylindric-ovate
Spruce, Serbian	Picea omorika	Cone	2-2 1/2	Glossy-brown
Spruce, Siberian	Picea obovata	Cone	2 1/2	Oblong-ovate, light brown
Spruce, Silver	Picea pungens 'Argentea'	Cone		

Spruce, Sitka	Picea sitchensis	Cone	2-4	
Spruce, Tigertail	Picea polita	Cone	4-5	Oblong, brown, glossy
Spruce, White	Picea glauca	Cone	1 1/2-2	Cylindric-oblong, light brown
Spruce, Wilson	Picea wilsonii	Cone		
Spruce, Yeddo	Picea jezoensis	Cone		
Spruce Pine	Pinus echinata	Cone	1 1/2-2 1/2	Conic-oblong, dull brown
Staggerbush	Lyonia mariana	Capsule	1/4	Pointed
Staghorn Sumac	Rhus typhina	Drupe	1 1/8 cluster	Persistent; round, red, sticky compact clusters, hairy
Star Acacia	Acacia verticillata	Pod	3	
Star Magnolia	Magnolia stellata	"Pod"	2	Red
Star Tulip	Calochortus species	Pod		Attractive when green, or when tan
Starry Campion	Silene stellata	Pod		
Starry Grasswort	Cerastium arvense	Capsule		As long as, or longer than, the calyx
Steeplebush	Spiraea tomentosa	Pod		Small, usually in groups

Common Name	Botanical Name	Type of Seed Vessel	Vessel Dimensions (length x width) in inches	Comments
Steeplebush (cont.) (Hardhack)				of 5; persistent
Sterculia, Hazel	Sterculia foetida	Pod	4	
Stereospermum	Stereospermum suaveolens	Capsule	18	3/4" across, woody
Stewartia	Stewartia pseudo-camellia	Capsule	5/8	Woody, 5-angled
Stillingia	Stillingia aquatica	Capsule	3/8	3-parted
Stinking Cedar	Torreya taxifolia	Drupe-like	1-1 1/2	Green-to-purple envelope
Stock	Matthiola incana	Pod		
Straggling St. John'-swort	Hypericum dolabriforme	Capsule	1/4	Egg-shaped, pointed
Strawberry Bush (Wahoo)	Euonymus americanus	Capsule	1/2-3/4 across	Rough-warty
Strawberry Bush, Running	Euonymus obovatus	Capsule	1/2 across	Rough-warty; pale orange-red
Strawberry Bush, Smooth	Euonymus americanus	Capsule	1/2-3/4 across	Rough-warty; crimson

Common Name	Scientific Name	Fruit	Size	Description
Strawberry Shrub	Calycanthus floridus	Capsule	2-2 1/2	Urn-shaped
Subalpine Fir	Abies lasiocarpa	Cone	2-4	Dark purple
Sugar Pine	Pinus lambertiana	Cone	10-20	Cylindrical, often slightly curved, light brown, lustrous
Sumac, Dwarf	Rhus copallina	Drupe	1 1/8 cluster	Persistent, round, red, hairy; large clusters
Sumac, Fragrant	Rhus aromatica	Drupe	1 1/4 cluster	Persistent, round, red, densely hairy; clusters
Sumac, Smooth	Rhus glabra	Drupe	1 1/8 cluster	Persistent, round, red, hairy; compact clusters, sticky
Sumac, Staghorn	Rhus typhina	Drupe	1 1/8 cluster	Persistent, round, red, hairy; compact clusters, sticky
Summer Hyacinth (Galtonia)	Galtonia candicans	Pod		
Sundrops	Oenothera fruticosa	Pod		Strongly ribbed
Supplejack	Berchemia scandens	Capsule	1/4	Oval, bluish-black
Swainsonia	Swainsonia galegifolia	Pod	1-2	Inflated
Swallow-wort, Black	Cynanchum nigrum	Pod	2	

Common Name	Botanical Name	Type of Seed Vessel	Vessel Dimensions (length x width) in inches	Comments
Swamp Azalea	Azalea viscosum	Capsule	1/2-3/4	Glandular, bristly, narrowly egg-shaped
Swamp Cyrilla	Cyrilla racemiflora	Capsule		Small, yellowish-brown, persistent, 4 seeded
Swamp Leucothoë	Leucothoë racemosa	Capsule	3/16 diameter	
Swamp Loosestrife	Decodon verticillatus	Capsule	1/4 across	Urn-shaped
Swamp Milkweed	Asclepias incarnata	Pod	3-4	
Swampbay Magnolia (Sweetbay ")	Magnolia virginiana	"Pod"	2-5	Fruits born in cone-like clusters; the individual fruits are pod-like, red
Sweet Cicely	Osmorhiza claytonii	Pod		
Sweet Gum	Liquidambar styraciflua	Ball-like	1-1 1/2	Bristly capsules; brown
Sweet Hakea	Hakea suaveolens	Pod		
Sweet Pea	Lathyrus odoratus	Pod		
Sweet Pepper Bush	Clethra alnifolia	Capsule	1/8	
Sweetbay Magnolia (Swampbay ")	Magnolia virginiana	"Pod"	2-5	Fruits borne in cone-like clusters; the individual fruits are pod-like, red

Sweetfern	Comptonia peregrina	Bur-like		Nutlets borne in little bur-like heads
Sweetgale	Myrica gale	Cone-like		Clusters
Sweetshrub, California	Calycanthus occidentalis	Pod	1 x 3/4	
Sweetshrub, Common (Carolina Allspice)	Calycanthus floridus	Capsule	2-2 1/2	Urn-shaped
Sweetshrub, Pale	Calycanthus fertilis	Pod	1 x 3/4	
Swiss Mountain Pine	Pinus mugo	Cone	2	
Swiss Stone Pine	Pinus cembra	Cone	2 1/2-3 1/2	Ovate, obtuse, light-brown
Sycamore (Buttonwood) (American Plane Tree)	Platanus occidentalis	"Ball"	1 x 1	
Table Mountain Pine	Pinus pungens	Cone	2 1/2-3 1/2	Conic-ovate, oblique at base, light brown
Tallow Tree, Chinese	Sapium sebiferum	Capsule	1/2 wide	
Tamarack (American Larch) (Eastern ")	Larix laricina	Cone	1/2-3/4	Oval or almost globular
Tamarind	Tamarindus indica	Pod	3-8 x 3/4-1	Velvety, scalloped, rust-brown

Common Name	Botanical Name	Type of Seed Vessel	Vessel Dimensions (length x width) in inches	Comments
Tamarisk	Tamarix species	Capsule	Very small	"Spikes"
Tanbark-Oak (Tanoak)	Lithocarpus densiflorus	Acorn		Enclosed in spiny, cup-like bracts
Tanoak (Tanbark-Oak)	Lithocarpus densiflorus	Acorn		Enclosed in spiny, cup-like bracts
Tansy, Common	Tanacetum vulgare	Pod		
Tar-Flower	Befaria racemosa	Capsule	1/4 x 1/4	
Tea	Thea sinensis	Capsule		Large
Tea, New Jersey	Ceanothus americanus	Capsule		Small, cup-shaped, persistent
Tea Tree, Australian	Leptospermum laevigatum	Pod		
Teasel	Dipsacus sylvestris	"Pod"	2-3 x 1	Spiny fruiting head
Teasel, Fuller's	Dipsacus fullonum	"Pod"	2-3 x 1	Spiny fruiting head
Tephrosia, Virginia	Tephrosia virginiana	Pod	1-2	Slender, silky
Thapsia	Thapsia villosa	Pod		
Thompson Magnolia	Magnolia x thompsoniana	"Pod"		

Tiger's Claw (Indian Coral-Bean)	Erythrina variegata orientalis	Pod		Black
Tigertail Spruce	Picea polita	Cone	4-5	Oblong, brown, glossy
Tiputree, Common	Tipuana tipu	Pod	2 1/2	
Tithonia	Tithonia rotundifolia	Capsule		
Toona	Toona febrifuga	Pod		
Tornillo (Screwbean)	Prosopis pubescens	Pod	1	Thin, glossy, light-tan, twisted like a corkscrew
Torrey Pine	Pinus torreyana	Cone	4-6	Broadly ovate, chocolate-brown
Torrey Vauquelinia	Vauquelinia californica	Capsule	1/4	Ovoid, woody, persistent
Torreya, California (Nutmeg, California)	Torreya californica	Drupe-like	1-1 1/2	Green-to-purple envelope
Tower Mustard	Arabis glabra	Pod		Long, bushy
Trachelospermum	Trachelospermum difforme	Pod	5-9	Slender
Trailing Arbutus	Epigaea repens	Capsule	1/4 x 1/4	
Traveler's Tree	Ravenala madagascariensis	Pod		Cluster; blue seeds in opened pod
Tree Peony	Paeonia suffruticosa	Pod		

Common Name	Botanical Name	Type of Seed Vessel	Vessel Dimensions (length x width) in inches	Comments
Tree-of-Heaven	Ailanthus altissima	Pod		Winged; rusty-red, on female trees
Trumpet Creeper (Trumpet Vine)	Campsis radicans	Pod	4-8	Slender, ridged
Trumpet Flower	Tecoma stans	Capsule	4-8	Linear
Trumpet Vine (Trumpet Creeper)	Campsis radicans	Pod	4-8	Slender, ridged
Tulip, Star	Calochortus species	Pod		Attractive when green, or when tan
Tulip Poplar (Yellow Poplar) (Tulip Tree)	Liriodendron tulipifera	Pod	2 x 3/4	Persistent
Tulip Tree (Tulip Poplar) (Yellow Poplar)	Liriodendron tulipifera	Pod	2 x 3/4	Persistent
Tutsan St. John'swort	Hypericum androsaemum	Capsule		Red at one stage
Umbrella Magnolia	Magnolia tripetala	"Pod"		
Umbrella Pine	Sciadopitys verticillata	Cone	3-5	

Common Name	Scientific Name	Type	Size	Description
Umbrella Tree (China-Berry)	Melia azedarach	Berry		Green in early stage; later, golden yellow; then rich brown
Unicorn Plant	Martynia louisianica	Capsule	4-6	Long, curving beak-odd shaped
Utah Juniper	Juniperus osteosperma	Cone	1/4-3/4	Dry, berry-like
Veitch Fir	Abies veitchii	Cone	1 1/2-4	Cylindrical, slender, dark purple
Veitch Magnolia	Magnolia x veitchii	"Pod"		
Veitch Sophora	Sophora davidii	Pod		
Velvet Leaf	Abutilon theophrastii	Pod	1	Circular clusters of 12-15 beaked pods
Vermillion Nasturtium	Tropaeolum speciosum	Capsule		Dull red
Veronica	Veronica incana	Capsule		
Vine-Wicky	Pieris phillyreifolia	Capsule	3/16 x 3/10	
Virginia Pine (Scrub Pine)	Pinus virginiana	Cone	1 1/2-2 1/2	Conic-oblong, reddish-brown
Virginia Spiraea	Spiraea virginiana	Pod		Small, usually in groups of 5; persistent
Virginia Stock	Malcomia maritima	Pod		

Common Name	Botanical Name	Type of Seed Vessel	Vessel Dimensions (length x width) in inches	Comments
Virginia Tephrosia	Tephrosia virginiana	Pod	1-2	Slender, silky
Virgin's Bower, Purple	Clematis verticillaris	Capsule	2	Plume-like tails - grayish-brown
Wafer Ash (Hoptree)	Ptelea trifoliata	Capsule	3/4	Papery; greenish; broad, thin, almost orbicular wing; drooping clusters up to 4 1/2" long; persistent
Wahoo (Strawberry Bush)	Euonymus americanus	Capsule	1/2-3/4	Rough, warty
Wallflower, Coast	Erysimum capitatum	Capsule	Up to 4	
Walnut, Black	Juglans nigra	Nut	2	Nearly round, sculptured
Washoe Pine	Pinus washoensis	Cone	9-12	
Waterer Laburnum	Laburnum x watereri	Pod		Pea-like
Watson Magnolia	Magnolia x watsonii	"Pod"		
Weather Plant	Abrus precatorius	Pod		
West Indies Mahogany	Swietenia mahogani	Capsule	3-4	Woody
Western Hemlock	Tsuga heterophylla	Cone	3/4-1	Oblong-ovoid

Western Larch	Larix occidentalis	Cone	1-1 1/2	Oblong
Western Red Cedar (Giant Arborvitae)	Thuja plicata	Cone	1/2	Clusters
Western White Pine	Pinus monticola	Cone	4-11	Cylindric, slender, slightly curved, yellowish brown
Western Yellow Pine	Pinus ponderosa	Cone	2 1/2-6	Ovate-oblong, light reddish or yellowish brown, and lustrous
Wheel-Stamen Tree	Trochodendron aralioides	Pod		Clusters of 5 to 10
White Ash	Fraxinus americana	Nutlet		Set in oblong, narrow "wing"; clusters
White Fir	Abies concolor	Cone	2-5	Oblong, gray-green, dark purple, or bright canary-yellow
White Lily Tree (Lily-of-the-Valley Tree)	Crinodendron dependens	Pod		
White Mustard	Brassica hirta	Pod	Up to 1 1/2	Bristling, ends in a flattened beak
White Pine	Pinus strobus	Cone	4-7	Cylindric, slender, oftened curved, reddish brown

Common Name	Botanical Name	Type of Seed Vessel	Vessel Dimensions (length x width) in inches	Comments
White Popinac (Leadtree)	Leucaena glauca	Pod	6	Narrow, reddish
White Poplar	Populus alba	Pod	Small	
White Spruce	Picea glauca	Cone	1 1/2-2	Cylindric-oblong, light brown
White Wicky	Kalmia cuneata	Capsule	1/8 x 1/8	
Whitebark Pine	Pinus albicaulis	Cone	4-6	
Whiteleaf Japanese Magnolia	Magnolia obovata	"Pod"		
Whitlow Grass	Draba verna	Pod		Oblong
Wicky, White	Kalmia cuneata	Capsule	1/8 x 1/8	
Wild Bergamot (Oswego Tea) (Bee Balm)	Monarda fistulosa	Capsule		Flower-like
Wild Ginger	Asarum canadense	Pod		
Wild Indigo	Baptisia tinctoria	Pod		
Wild Mock-Cucumber	Echinocystis lobata	Pod	2	Papery, puffed

Wild Mustard (Field Mustard)	Brassica kaber pinnatifida	Pod	3/4	Conspicuously beaded
Wild Radish	Raphanus raphanistrum	Pod		
Wild Senna	Cassia marilandica	Pod	4	
Wiliwili Tree	Erythrina sandwicensis	Pod		Narrow, curling; red seeds
Willow, Desert	Chilopsis linearis	Capsule	7-12 x 1/3 thick	Persistent
Willow, False	Baccharis angustifolia	Capsule		Small, clusters
Willow, Peachleaf	Salix amygdaloides	Capsule	1/4	Globose-conic; clusters about 2" long
Willow Myrtle (Peppermint Tree)	Agonis flexuosa	Capsule		Woody
Wilson Magnolia	Magnolia wilsonii	"Pod"		
Wilson Spruce	Picea wilsonii	Cone		
Wind Poppy (Flaming Poppy)	Meconopsis heterophylla	Capsule		
Winged Spindle Tree	Euonymus alatus	Pod		Usually in 4's; showy; bright-scarlet interior; persistent
Winter Aconite	Eranthis hyemalis	Pod		

Common Name	Botanical Name	Type of Seed Vessel	Vessel Dimensions (length x width) in inches	Comments
Winter Cress	Barbarea vulgaris	Pod		Erect
Wintergreen, Spotted	Chimaphila maculata	Capsule	1/4	Round
Winter-Hazel	Corylopsis species	Capsule		Woody
Winter-Purslane	Montia perfoliata	Pod		
Wisteria, American	Wisteria frutescens	Pod	2-4	Bean-like, knobby; fuzzy
Wisteria, Chinese	Wisteria sinensis	Pod	4 1/2-7	Velvety; hanging
Wisteria, Japanese	Wisteria floribunda	Pod	4 1/2-7	Velvety, hanging
Wisteria, Kentucky	Wisteria macrostachya	Pod	2 1/2-5	Bean-like, knobby, fuzzy
Witch-Alder, Dwarf	Fothergilla gardenii	Capsule	Less than 1/2	Egg-shaped, woody, downy, 2-beaked
Witch-Alder, Mountain (Fothergilla, Large)	Fothergilla major	Capsule	1/2	Egg-shaped, woody, downy, 2-beaked
Witch-Hazel	Hamamelis species	Capsule		Woody, urn-shaped
Witch-Hazel, Common	Hamamelis virginiana	Capsule		Woody, urn-shaped, grayish-downy
Woad	Isatis tinctoria	Pod		Dark-brown, oblong or orbicular

Woman's-Tongue Tree (Lebbeck)	Albizzia lebbek	Pod	12	Flat, lustrous
Wonga-Wonga Vine	Pandorea australis	Pod		
Wood Rose	Ipomoea tuberosa	Pod-like		Brown, globular, with dried and hardened sepals
Wooden Cherries	Argyreia nervosa	Pod		
Woolly Pipevine	Aristolochia tomentosa	Capsule	2-3	Ribbed, woolly
Woolly Sweet Pepper-bush	Clethra alnifolia var. tomentosa	Capsule	1/8	Roundish
Xanthoceras	Xanthoceras sorbifolium	Pod	2-3	Greenish-bur-like
Yam Bean	Pachyrhizus erosus	Pod	6-9	
Yeddo Spruce	Picea jezoensis	Cone		
Yellow Bells	Stenolobium stans	Pod	6-8	Narrow
Yellow Buckeye	Aesculus octandra	Capsule	2-3	Smooth
Yellow Cress	Rorippa islandica	Pod		Short, roundish
Yellow Cress	Rorippa sinuata	Pod		Strongly curved, hook-like
Yellow Cress	Rorippa sylvestris	Pod	1/2-2/3	Slender, slightly curved

Common Name	Botanical Name	Type of Seed Vessel	Vessel Dimensions (length x width) in inches	Comments
Yellow Cucumber Tree	Magnolia cordata	"Pod"	2 1/2	
Yellow Elder	Tecoma stans	Capsule	5-7	
Yellow Flag Iris	Iris pseudacorus	Pod		
Yellow Flax	Linum virginianum	Pod		Small
Yellow Jessamine (Carolinia Jessamine) (False Jessamine)	Gelsemium sempervirens	Pod	1/2-3/4	Egg-shaped, flattened, short-beaked
Yellow Poplar (Tulip Poplar) (Tulip Tree)	Liriodendron tulipfera	Pod	2 x 3/4	
Yellow Star Jessamine	Trachelospermum asiaticum	Pod		
Yellow Turk's Cap Lily	Lilium pyrenaicum	Pod		
Yellowhorn, Shinyleaf	Xanthoceras sorbifolium	Bur		
Yellowwood	Cladastris lutea	Pod	4-5	Irregularly shaped, flat
Yerba Mansa	Anemopsis californica	Pod		
Yew, Plum	Cephalotaxus drupacea	Drupe-like	1	Oval; green

Yucca	Yucca whipplei	Capsule		
Yucca, Mound-lily	Yucca gloriosa	Capsule	2-3	
Yucca, Soaptree (Spanish-Bayonet)	Yucca elata	Capsule	1 1/2-2	Oblong, thin, woody, light brown
Yulan Magnolia	Magnolia denudata	"Pod"	3-4	Brownish, cylinder-like
Zenobia	Zenobia pulverulenta	Capsule	1/4	Roundish
Zylomelum	Zylomelum angustifolia	Pod		Fish-like shape

CONTEMPORARY PRACTICES INDEX

INDEX TO BLEACHED ELEMENTS

BOTANICAL NAMES

INDEX TO GRASSES AND GRAINS

BOTANICAL NAMES

INDEX TO PODS, CONES, AND OTHER SEED VESSELS

BOTANICAL NAMES